ALEXANDER HUNTER

A TREATISE OF WEIGHTS, METS AND MEASURES OF SCOTLAND

EDINBURGH, 1624

WALTER J. JOHNSON, INC.
THEATRUM ORBIS TERRARUM, LTD.
AMSTERDAM 1974 NORWOOD, N.J.

The publishers acknowledge their gratitude to
the Trustees of the British Museum for their
permission to reproduce the Library's copy,
Shelfmark: C.38.d.12

S.T.C. No. 13993

Collation: π^4, A-F^4, G-I^2, K^4

Published in 1974 by

Theatrum Orbis Terrarum, Ltd.
O.Z. Voorburgwal 85, Amsterdam

&

Walter J. Johnson, Inc.
355 Chestnut Street
Norwood, New Jersey
07648

Printed in the Netherlands

ISBN 90 221 0671 3

Library of Congress Catalog Card Number:
74-80191

NUMBER 671

THE ENGLISH EXPERIENCE

ITS RECORD IN EARLY PRINTED BOOKS
PUBLISHED IN FACSIMILE

A TREATISE,
OF WEIGHTS, METS AND MEASVRES OF SCOTLAND.

WITH THEIR QVANTITIES, AND TRVE Foundation, and sundrie profitable Observations, arising vpon everie one of them, Together with the Art of Metting, measuring & compting all sort of land with diverse Tables.

BY ALEXANDER HVNTAR,
BVRGES OF EDINBVRGH.

PROVERBS, chap. 20. verse, 10.
Diverse Weigthes, and diverse Measures, both these are abhomination vnto the LORD.

EDINBVRGH,
Printed by *Iohn Wreittoun*, and are to bee solde at his *Buith*, at the *Nether-Bowe*.
WITH LICENCE

Per Regem.

THE Lords of his M. secret Counsell expresselie inhibiteth and dischargeth all persons whatsoever, to print, or cause to bee printed, sell, or cause to bee solde: this Treatise anent the Weights, Metts, and Measures of SCOTLAND with the Arte of Metting, Measuring, and Compting of Land, with diverse Tables, both in large, and briefe manner, and other thinges, composed by ALAEXANDER HVNTAR Burgesse of Edinburgh, for the space of Ten yeeres after the date heereof, without the consent of the said ALEXANDER HVNTAR his Heires, and assignes ynder the paine of escheating the Bookes, and Papers to the said ALEXANDER and his foresaids vse. Given at EDINBBVRGH, the third day of Februar. Anno Dom. 1624.

IACOBVS PRYMEROSE.

TO THE RIGHT HONORABLE,
and his very good Lord,
S^r.
GEORGE HAY, OF KINFAVVNES,
CHIEFE CHANCELOR
OF SCOTLAND.

HERE are three thinges (RIGHT HONORABLE) moving and imboldening mee to dedicate to your L. this litle Treatise. The first is, the singulare good-will, and vndeserved favour I haue found at all times at your L. hands. The second is, the great zeale, and affection you beare as a Father to nourish Learning, honest endeavours

vours, and vertuous exercises profitable for this Kingdome, and Common-wealth, as testifie the Workes erected, and maintained those manie yeeres, which are to your L. no smal charges, but greater honour and fame. The third and last is, this beeing a Treatise of Compting your L. can best judge of it, who hath given sufficient proofe thereof, for the benefite aswell of our neighbour-Countrie, as our owne. These are causes moving mee to presente to your L, the fruite of my Travelles. Hoping that your L. will accept of it, as a Token of my humble duetie, and simple remembrance: That it comming to light vnder the shelter of your wings, may bee protected from the envious, and with the readier minde receaved of the vertuous. So wishing your L. increase, grace, and happinesse, I rest,

Your L. most humble affectionat

ALEXANDER HVNTAR.

TO THE READER.

I Haue set downe here (Gentle Reader,) in vulgar tearmes for the benefite of all, a necessarie Treatise devided into three partes. The first part, concerneth the Weightes, with the Metts both liquid and drie, and the Measures of Scotland, describing their just quantities, with the foundation whereuppon everie one of them is grounded, with divers observations arising vppon everie one of them in particular, not knowne to many and yet needfull to bee vnderstood of all. Secondly, you haue the art of Metting and Measuring of all kinde of land grounded vpon the said Measures: showing how to reduce and bring vnequall pieces of land in sundrie formes and fashions that they may bee the better measured, with the manner of measuring every fashion of land by it selfe particularly: and thereafter to finde speedelie without compting the just quantitie of everie piece of land in Acres, Roodes, Fallis, and Ells, by a large Table made to that effect. Thridly, because there is divers workes that are Mett and measured both in length and breadth, as pieces of Tapistrie, Sclaiting of houses, building of Walles and Dykes, with the Glazen-wrightes worke and such like: the manner of the compt and reckening whereof is not knowne to many, Therefore there is another Table also made to help the ignorant compter, whereby the

most

THE PREFACE.

moſt ſimple ſhall inſtantly finde the juſt compt of every ſort of worke being juſtlie meaſured, as at more length is ſet downe in the deſcription of the ſaid Table. Farther there is a Table needefull for all indwellers within Burgh, ſhewing what everie loafe of Wheate bread ſhould weigh at all prices of wheate, conforme to the declaration thereof.

In ſetting downe hereof, I can not giue contentment to all, becauſe ſome will finde fault and diſpraiſe that which others will eſteeme of, and others will ſay they could haue done this much better then I haue done it: I grant there is many whoſe skill is better then mine, if they could take the paines: but ſeeing they are ſlack in ſo needefull a worke, let them not be offended with me in preventing them. I doe not preſume to profite ſuch as vnderſtandes, but the ſimple and vulgar ſorte, who hath not heard of the like after our Scottiſh reckening: I hope that this groſſe beginning ſhall encourage ſome of better vnderſtanding to write farther vppon this ſubject, or to ſet foorth the like for the benefite of their countrie. And in the meane time I will requeſt you, who hath a deſire to profite by theſe my ſmall labours, that you will reade it through of purpoſe, rather to vnderſtand then to carp at it: conſider it ſoundly, and you will finde both the practiſe eaſie and all thinges plaine: and where any parte ſeemeth to bee obſcure, I wiſh my ſelfe to bee preſent to reſolve you. Accept therefore my honeſt intention in good part, and if I heare that it bee receaved without detracting, it will be a meanes to encourage me to a farther labour for your benefite. And ſo I reſt. Fare-well.

THE

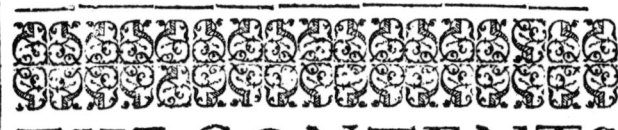

THE CONTENTS
OF THIS TREATISE FOLLOWING.

THE FIRST PART.

	Folio
OF *Weightes and the foundation thereof.*	2
Of liquid Metts and their foundation.	4
Of drye Metts.	5
Of Measures for length and breadth.	5
Of the Roode of worke.	6
The foundation of the Scottish Acre of Land.	6
Of the English Acre.	7
The foundation of the Myle of ground.	8
The proportion and difference betwixt Measures.	9
Of the English Myle.	10
Of the Scottish Myle.	10

THE SECOND PART.

Of the Instrument, wherewith Land is measured.	12
The vse of the said Instrument.	13
Description of the Table made to finde the compt of measured Land.	13
How Quadrangles and Square Land should bee measured and compted.	17
To measure Land that is of vnequall breadth.	18
Of Land that is vnequall both in length and breadth.	19
Of Triangles.	20

Of

THE CONTENTS.

Of the discending lyne in Triangles.	21
Of vnequall pieces of Land and counting thereof.	23
To devide vnequall pieces of Land.	24
To measure Circles or Land that is round.	27
To measure Triangled Squares.	28
To measure Woodes or deformed Lands.	29
To measure Mountaines and Hilles.	31
To measure Valleis or low ground.	33
To finde the compt when small measures happeneth in the breadth of every Land.	35
When small measures happeneth in the length.	35
Description of the small Table.	36
When small measures happeneth both in the length and in the breadth.	38
The beginning of the great Table.	41

THE THRID PART.

Of Building and Sclaiting and counting thereof.	55
Description of the Table of Sclaiting.	55
Example of the Masons worke.	56
Example of Sclaiting and counting thereof.	56
Example of Tapistrie and counting thereof.	57
The small Table to that effect.	58
The beginning of the Table of Sclaiting.	59
The description of the Table of wheate bread.	63
Example how to finde the weight of wheate bread.	64
The beginning of the Table thereof.	66
The Conclusion.	68

FINIS.

THE FIRST PART OF THIS TREATISE CON-
CERNETH THE DESCRIPTION
OF THE WEIGHTES, METTS,
AND MEASVRES OF THIS
KINGDOME, AND OF
certaine Documentes,
arising therevppon.

AS at the beginning all thinges were disposed and made in Measure, Number, and Weight: So for Policie, and good Governement, in this kingdome of Scotland like vnto other Nations, wee haue our severall Weightes, Metts, and Measures: with the foundation wherevppon they were made and ordained. And in the first, wee haue our Weights at the ounce weight, the Pound, and the Stone weight, for weying of Gold, Silver, Silke, Sugar, Spyces, Bread, Wooll, Butter, Cheese, Iron, and other commodities, vppon the which weight is founded our liquid Metts, as the Choppin, the Pinte, Quart, and Gallon, for metting of Wine, Ale, Beere, Vineger, Oyle, Aqua-vitie, with the like liquid commodities: So wee haue our drie Metts, as the Peck, the Firlet, and the Bow, for metting of al Cornes, Salt, Coales, and other drie commodities. In like manner, wee haue our Mea-

A   sures,

sures, for length, breadth and thicknes, as the Inch, the Foote, the Faddome, and the Ell: which are the foundation to measure all Marchandise with the Rood of worke, the Aiker of land, and partes thereof, with the mile of ground and quantitie thereof, all set downe at length hereafter, And first.

OF THE WEIGHTS,
and foundation thereof.

A corne or pickle of wheat, taken out of the middest of an eare of wheate, is the foundation of a graine weight.

18. of those graines, maketh the halfe drop weight.

36. graines maketh a drop weight.

4. drop weight, is a quarter of an vnce.

4. quarters, maketh an vnce weight.

8. vnces is a mark weight.

2. marke weight is a pund weight.

16. pund weight maketh the stone weight of Lanerke.

The Standard Stone of Lanerk.

There was also a Trone stone weight, which did wey 19. pundes and 8 vnces of *Parise* weight, wherewith the Butter, Cheese, Wooll, Tallow, and such other Countrie commodities as carryeth refuse was weyed.

There is other quantities whereof the weight is here set downe, to giue some contentment to the Reader, as a Tunne weight of light goodes, which is the common fraughting of all Marchandice betwixt this Countrie, and France, England, or Spaine, which Tunne is esteemed to wey 600 pund weight.

A sack of goodes which is commonly the fraughting of all Marchandice from this countrey, to the Low

Low-Countries, esteemed to weye 40 stones, or 640 pund weight.

The Sirplithe of goodes, which is the common fraughting of Marchandice betwixt this Countrey, and the Easterne Countreyes, is esteemed, to wey 80. stone weight, or 1280 pund weight.

The Last of goodes, is esteemed to wey 120 stone weight or 1920 pund weight.

The Fidder of Leade, is 126 stone of 2000 pund weight.

THE WEIGHTS FOLLOWING,
are vsed by Apothecaries, in mixture of *their Medicines wherein the least is a graine.*

24. graines maketh a Scruple.
3. Scruples maketh a Dragme.
8. Dragmes maketh an vnce.
16. vnces maketh a pund.

A Shekle weight mentioned in the Bible, is halfe an vnce.

A Talent is compted to 120 pund weight.

A Talent of money is 600. Crownes.

The weight of all quantities of Wheat bread, at everie price of Wheat, is set downe in a Table heere after following.

In all our neighbour-Countries the Fleshe is solde by weight.

If the Meale were solde also by weighte, it might prove proffitable to the Lieges.

The 12 ounces Troy weight of England, weyes 12 ounces 3 drop weight 21 graines Scottish weight.

Now of these our weightes are made other measures both for cornes and liquid stuffe.

Of Weights, Metts, and Measures.

THE FOVNDATION OF THE liquid Metts proceeding from the Weight.

The standerd Iuge or pinte of Striueling.

The Scottish pinte or standerd Iug of Sterling, is found to conteine 3 pound 7 ounce Weight of the water of Leith, everie pinte is devided in 2 choppins and 4 muchkins.

2. Pintes maketh a quart.
4. Quartes or 8 pintes is a Gallon.
The Salmon barrell conteines 10 Gallons.
The Herring barrell holdes 8 Gallons and a halfe.
2. Barrells or 17 Gallons is the full of a Burdeaux Puncheon.
The Puhcheons of high countrey Wines are of 13 or 14 gallons.
4. Puncheons makes a Tunne.
2. Pipes is a Tunne.
6. Barrekines makes a Tunne.
6. English bunes of Beere is a Tunne.
6. Salmon barrells is a Tunne.
8. Herring barrells is a Tunne.
12. Barrells makes a Last.

Our Scottish pinte conteines very neere 4 pintes and a halfe of English.

Now if the ground be trew that the pinte doth weigh 55 ounces, then consequentlie.

The Tunne should weigh,	116. stone 14 pound.
The Puncheon full,	29. stone 3. pound 8 vnces.
The barrell being full,	14. stone 9. pound 12 vnces.
The gallon should weigh,	1. stone 11. pound 8 vnces.
The quart full,	6. pound 14 vnces.
The pinte,	3. pound 7 vnces.
The choppine,	1. pound 11 vnces and a half.
The muchkin full,	13. vnces 12 drop.

As of weights did spring these liquid Metts, so of the same ariseth the drie Metts, as Pecks, Furlets, and Bowes.

THE FOVNDATION OF THE drie Metts.

The Firlet of Linlithgow, which is the standerd for the whole countrie, for metting of Wheate, Rye, Beans, Pease, Meale, or white Salt, conteines 21 pintes and a Muchkin of the water of Leith: which Firlet is devided in 4 Peckes. with halfe peck, and fourth part of the Peck. *[marginal: The Standerd Firlet of Linlithgow.]*

The Firlet (for metting of Beere, Malt, or Oates, which were called heaped cornes,) conteines 31 pintes of water.

4. Firlets makes the Bow.
16. Bowes is a Chalder.
18. Bowes and a halfe is compted for a last of Rye.

The halfe bowe mett of the water measure of Lieth conteines 9 peckes.

The English quarter of corne, conteines hard by 2. bowes of Scottish measure.

The Bow of wheate will weigh,   14. stone 3 pound.
The Firlet of drie Wheate, 3. stone 8 pound 12 vnces.
The peck will weigh,   14. pound 3 vnces.
The halfe peck,   7. pound 1 vnce and a halfe.

THE FOVNDATION OF MEAsures, for length, breadth, and thicknes.

3. Barlie cornes faire and round lying in length without the tailes maketh an inch.
12. Inches maketh a foote.
3. Foote is an English yard.
3. Foote and an inch, or 37 inches makes the Ell of Edinburgh. Which Ell is parted in 4 quarters, and everie quarter in 4 nailes. *[marginal: The Standerd Ell of Edinburgh.]*

45 In-

45. Inches is the English Ell.
27. Inches is the Flemish Ell.

In France everie Towne hath a diuers measure.

The foote wherewith the glazen-wrightes measures their worke, some are of 9 inches, and some but 8 inches.

In our neighbour countries, and West part of this countrie, all timber is sold by measuring the length, the breadth, and thicknes thereof, and compted by the foote of square. To knowe what number of square feete or other measures everie piece of measured Timber containes, it may be helped by a Table, if it be found needfull.

OF THE ROOD OF WORKE.

A Rood of land containes 240 Ells of measure: But a Roode of worke, wrought by Masons or Sclaiters, containes but 36 Ells: that is, if any piece of worke bee found to bee 18 Ells in length, and 2 Ells in breadth, it makes a Roode.

12. Ells in length, and 3 Ells in breadth is a Roode.
9. Ells in length, and 4 in breadth is a Roode.
8. Ells in length, and 4 Ells and a halfe in breadth is a Roode.
6. Ells in length, and 6 Ells in breadth is a Roode.

A Roode of land within Burgh, is esteemed of old to bee 20 foote: that is, 5 foote in length, and 4 foote in breadth.

THE FOVNDATION OF AN
Acre of land.

6. Ells of the standerd of Edinburgh, makes a lineall fall

Of Measures.

fall, wherewith land is measured.

6: Ells long, and 6 Ells breadth, makes a superficiall or square fall, wherewith land is reckened.

40. Falles makes a Roode.
10. Falles in length, and 4 in breadth, is a Roode.
8. Falles in length, and 5 in breadth, is a Roode.
4. Roodes is an Acre.
So an Acre containes 160 Falles, or 960 Ells.
80. Falles in length, and 2 Falles in breadth, maketh an Acre.
40. Falles in length and 4. in breadth maketh an Acre.
32. Falles in length, and 5 Falles in breadth is an Acre.
20. Falles in length, and 8 in breadth, is an acre.
16. Falles in length, and 10. in breadth is an Acre.

4. Acres are compted for a Ministers Gleib.
6. Acres arable land, for an houseband land.
13. Acres is compted an Oxen-gate.
4. Oxen-gate is esteemed a pund land of old extent.

THE FOVNDATION
of an Englifh Acre.

3. Barliecornes maketh an inche.
12. Inches maketh a Foote.
3. Foote is an Englifh Yarde.
5. Yardes and a halfe maketh a pearch.
40. Pearches are a Rood.
4. Roodes are an Acre.
So an Englifh Acre is 160. pearches, or 880. of yards

pearches, which is of Scottish measure 856. Elnes and 8 inches.

By this accompt the Scottish Acre, is 103 Ells and 29 inches of Scottish measure more than the English Acre.

THE FOVNDATION OF MEASVRES, AND OF THE MYLE.

4. Cornes of barlie Beir, lying in breadth maketh a finger breadth.
4. Fingers breadth, is a palme.
3. Inches is a palme.
3. Palmes is a spanne.
4. Palmes is a foote.
6. Foote is a fathome.
6. Palmes is a cubite.
5. Foote is a pace.
125. Paces is a furlong.
8. Furlongs is an Italian Myle.
2. Myles is a French Liegue.
4. Myles is a Germane Myle.

HEERE

Of Measures. 9

HEERE FOLLOWETH THE PROPORTION AND DIFFE-
rence betwixt Measures, and what
number of everie small measures,
each great Measure doeth
containe.

A Scottish Myle containeth

1 of
- Furlongs —————— 8
- Falles ——————— 320
- Fathomes — 986 & 4 foot.
- Paces ——————— 1184
- Elnes ——————— 1920
- Cubites — 3946 & 1 foot.
- Footes ——————— 5920
- Spannes — 7893 & 3 inches
- Palmes ——————— 23680
- Inches ——————— 71040
- Fingers ——————— 94720

A Furlong containeth,

2 of
- Falles ——————— 40
- Fathomes — 123 & 2 foot.
- Paces ——————— 148
- Elnes ——————— 240
- Cubites — 493 & 6 incges.
- Footes ——————— 740
- Spannes — 986 & 6 inches
- Palmes ——————— 2960
- Inches ——————— 8880
- Fingers ——————— 11840

A Fall containeh,

3 of
- Fathomes — 3 & 6 inches
- Paces ——— 3 & 42 inches
- Elnes ——————— 6
- Cubites — 12 & 6 inches
- Footes — 18 & 6 inches
- Spannes — 24 & 6 inches
- Palmes ——————— 74
- Inches ——————— 222
- Fingers ——————— 296

A Fathome containeth

4 of
- Paces ——— 1 & 1 foot.
- Elnes ——— 1 & 3 5inches
- Cubites ——————— 4
- Footes ——————— 6
- Spannes ——————— 8
- Palmes ——————— 24
- Inches ——————— 72
- Fingers ——————— 96

A Pace containeth,

5 of
- Elnes ——— 1 & 23 inches
- Cubites — 3 & 6 inches
- Footes ——————— 5
- Spannes — 6 & 6 inches
- Palmes ——————— 20
- Inches ——————— 60
- Fingers ——————— 80

An Elne containeth,

6 of
- Cubites ——— 2 & 1 inche.
- Footes ——— 3 & 1 inche.
- Spannes ——— 4 & 1 inche
- Palmes ——— 12 & 1 inches.
- Inches ——————— 37
- Fingers ——— 49 & 3 parte.

A Cubite containeth,
7 of
{
Footes ——— 1 & 6 inches.
Spannes ——— 2
Palmes ——— 6
Inches ——— 18
Fingers ——— 24
}

A Foote, esteemed the 6. parte of a mans length containeth,
8 of
{
Spannes —— 1 & 3 inches.
Palmes ——— 4
Inches ——— 12
Fingers ——— 16
}

A Spanne containeth,
9 of
{
Palmes ——— 3
Inches ——— 9
Fingers ——— 12
}

A Palme containeth,
of
{
Inches ——— 3
Fingers ——— 4
}

Of the fundementall Myle which containeth, Of paces 1000. which is of English measure 1666 yards & 2 foots, and of Scottish measure 1621 Elnes 23 inches.

OF THE ENGLISH MYLE.

They compt 40 pearches to a furlong, and 8 furlongs to a myle which is 320 pearchs or 1760 yards, & containeth of Paces 1056. Of English measure 1760. yardes, and of Scottish measure 1712 Ells.

So the English Myle is more than the Fundementall, or Italian Myle of paces 56, of English yardes, 93 and 1 inch, and of Scottish measure 90, Elnes and 30, inches.

OF THE SCOTTISH MYLE.

40. Falles is a furlong, 8 furlongs is 1 myle, which is 320. Falles. It containeth of paces 1184, of English measure 1973 yards, 1920 Elne Scottish, & of Fathoms 986 and 2 foote. So the Scottish Myle is more than the Italian Myle 184 paces, of English measure 306 yards 2 foots, of Scottish measure 298 Ells and 13 inches. And it is more than the English Myle of paces 128. Of English measure 213 yards, 12 inches. And of Scottish measure 207 Elnes, and 20 inches.

A Square Scottish Myle, that is a myle of length and a myle of breadth, containeth 640. Acres of land.

THE
SECOND PART
OF THIS TREATISE
CONCERNETH THE METTING
AND MEASVRING OF LAND
FOVNDED VPON THE FOR-
MER MEASVRES.

ALBEIT there be many perſons in the countrie that profeſſeth to bee meaſurers of land,
and that ſundrie hath written vppon the meaſuring of land in divers languages : where you may
learne a great deale more then is here ſet downe. Yet
becauſe that ſome Heritoures of landes, will deſire to
haue their landes mett and meaſured to know the quantitie thereof for their pleaſure, when they can not haue
a land meaſurer to ſerue them, neither bookes to informe
them according to our Scottiſh meaſures. Therefore
to giue them contentment that they themſelues or ſer-
vants may meaſure all kinde of grounde: although it be
arable land, Mures, Medowes, Moſſes, Loches, Hills,
or valley ground, and knowe what everie piece thereof
doeth containe in quantitie. There is here ſet downe,
not onely the way how land ſhould bee meaſured : but
alſo how to finde the quantitie thereof. For albeit that
land bee meaſured both in length and breadth, that re-
ſolves not what number of Acres, Roodes, and other
ſmall quantities it containes, before the compt thereof
bee caſt by Arithmetique, and the length bee multiplyed

plyed by the breadth, and thereafter devided: and because there is not many that can multiplie and divide numbers, and that I haue seene great ignorance in some land measurers, in making of the cempt after the land was measured. Therefore to eschew negligent compting my cheife care is, to set downe a perfite and just Table: where you shall speedilie finde without compting the quantities that any land conteines after that the trew length and breadth is found out, as is at length set downe hereafter.

In the metting and measuring of ground: First wee should know the just length and breadth thereof, next what number of Acres, Roodes, and Falles arriseth vpon everie length and breadth. Now to finde the length and breadth, wee must know by what instrument it is found, and how to vse the same, and to finde what number of Acres ariseth vppon the length and breadth: the compt thereof must bee cast by Arithmetike, or found by the Table after following.

THE INSTRVMENT WHEREWITH land is measured.

The said instrument is knowne to bee two staues, everie one of them 6 quarters long or thereby pricked with iron, hauing the trew measures of an Ell, halfe Ell and quarter Ell marked vppon them, with a coard or small cheine the length of 6 Ells, made fast betweene the said staues, a shaft length aboue the prickes: which coard would be either barked or well seared with waxe or roset: Remembring alwayes in case you haue any great boundes of land to measure, then your coard or chaine would bee of 18 or 12 Ells long at the least.

THE

THE VSE OF THE SAID
Inſtrument.

The vſe thereof is, that 2 men ſhall carrie the ſaids ſtaues, and ſhall begin at the end of the land, hauing the ſaid coard ſtretched and ſtented to the full length betweene them, and with that meaſure euerie ſquare piece of land is meſured over in the middeſt, what Fallis and Ells it hath of length: and thereafter is meaſured croſſe over the middeſt, what Fallis and Ells it hath of breadth, and a note ſet downe in write of the juſt length and juſt breadth: Remembring that the breadth or wideneſſe ſhould bee truelie ſearched, becauſe a little errour in the breadth increaſeth to a great fault in the length. And thus much for the ſaid inſtrument and vſe thereof.

Before any examples are ſet downe for meaſuring of land, it is neceſſarie to ſet downe the deſcription of the Table, where to finde the compt of all land that ſhall happen to bee meaſured.

THE DESCRIPTION OF THE
Table, to finde the compt of
meaſured land.

There is none ſo ignorant, but they doe, or may eaſilie know, the names of theſe ten figures, 1. 2. 3. 4. 5. 6. 7. 8. 9. 0. with their ſtrength in the firſt and ſecond place, and by a little frequenting thereof, they may attaine to reade and vnderſtand this Table, and the reſt of the Tables following conforme to their deſcriptions. And firſt, this Table I haue made and comprehended all in the boundes of a ſheet of paper, but it is ſet downe here in an ample and large manner, to the intent that a part thereof may juſtifie the other, and that the common and vulgar ſort to Landwart may eaſily vnderſtand it, It is grounded vppon the Ell of meaſure: whereof

B 3   6 in

6 in length is compted for a Fall, 40 Fallis for a Roode, and 4 Roodes for an Acre, as is set downe before. It conteines sundrie diverse pages: In everie page there is 4 Columnes, and everie Columne thereof containeth 3 partes: To wit, the breadth of the land with the length thereof, and the quantitie of the number of Acres, Roodes, and Falles that riseth vppon everie severall length and breadth, The breadth of the land is set downe vppon the head of everie Columne, as vppon the first page there is the Columne of a quarter Ell, the Columne of a halfe Ell, with the Columne of three quarters of an Ell: And the Columne of an Ell vpon the second page, the Columnes of 2 Ells 3. 4. 5. Ells, which are the small measures: Then vppon the thrid page, beginneth the Columne of 1 Fall, of 2. 3. and 4, Fallis, and so foorth in order to 30 Fallis, and to 100 Fallis of breadth. The length is set downe vppon the left side of everie Columne, and goeth downe from the head to the foote of the page, betweene the two small lines, beginning at one Fall, to 25 Fallis, and to 200 Fallis. The product of the number of Acres, that riseth vpon the compt of the length and breadth, is set downe in the broade space of everie Columne, against the length in the narrowe: containing 3 numbers, titled and named vppon their heades, with Acres, Roodes, Falles, Ells, and quarter Ells. Now to finde the compt that any land extendes to being measured in length and breadth: you shall ever seeke the breadth vppon the head of the Table, and the length vppon the left side of that Columne, and in the broade roome against the length, you will finde the aunswer what the compt extendes to. Example, a piece of land is founde to bee 80 Fallis of length, and 17 Fallis in breadth, you shall

seeke

Of Measuring of land. 15

seeke the Columne of 17 Fallis vppon the head of the Table, and in that same Columne seeke the length 80 and you will finde againſt 80, to the right hand 8 Acres 2 Roodes, which is the quantitie thereof. Another example: A piece of land 70 Fallis of length, and 21 Fallis of breadth, seeke the Columne of 21 Fallis vppon the head of the Table, and then seeke the length 70. In the left side of that same Columne, and againſt it to the right hand, you will finde 9 Acres and 30 Fallis. But becauſe it may happen that some defect will bee in the printing of this Table and the next: or that any other occasion fall out, that you are not well reſolved of the quantitie of the compt: therefore to juſtifie the Table, and to giue yon contentment, you ſhall finde the compt reſolved three manner of wayes: The firſt is, to seeke the breadth of the land vppon the head of the Table, and the length vppon the side of the Table, as is set downe in the former examples: The second waye is to seeke it contrarie-wise, that is to seeke the breadth in the side of the Table, and the length vppon the head of the Table, and in the broad roome you will finde the same compt that the firſt produced. The third way to finde the compt is to devide the length in two or three partes, and to seeke the compt at sundrie times, as if the number of the length bee 24: to seeke firſt the compt of 20, and then the compt of 4: and if the length 18, to seeke firſt the compt of 10, and then the compt of 8, or seeke 9. 2 times, will bee 18, and you will finde that all these formes will yeeld alike compt. Example, A piece of land is found to bee 90 Fallis of length, and 24 Fallis of breadth, if you seeke the compt thereof after the firſt way, which is the easieſt and beſt way, you will finde in the Columne of 24

againſt

against 90, standing 13 Acres 2 Roodes: To seeke it after the second manner, you will finde in the Columne of 90 against 24, the same compt of 13 Acres and 2 Roodes: And to seeke it after the thrid forme, you shall cast the length 90 in 2 partes: to wit, in 40 and in 50, which maketh 90: or in 60 and 30 which maketh likewise 90, and you will finde in the Columne of 60 against 24. 9 Acres: and you will finde in the Columne of 30 against 24. 4 Acres 2 Roodes: These two being added together will yeeld the foresaid compt of 13 Acres and 2 Roodes, and so all the three formes will yeeld alike compt. The like forme of tryall may bee vsed with the other Table concerning Building and Sclaiting.

Now followeth the way to measure all sort of land, but before my examples are set downe touching it: you must consider, that there is divers fashions of land, and therefore to bee measured in divers manners: and some manner of land lieth in such sundrie formes, that it can not bee measured, but in divers partes: then consider how many partes, and in what manner of fashion they must bee devided, that you may n sure everie part according to their forme and fashion: and how so ever the piece of land bee formed or fashioned, bee it square, bee it round or triangle, mounting to a hill, or descending in a valley, it must bee reduced and brought to a certaine length and certaine breadth, otherwise it can not bee brought and summed to a perfite quantitie of Acres, and other odd quantities.

OF

OF THE RVLE
OF QVADRANGLES, AND HOW ALL SQVAIRE LAND SHOVLD BE MEA-SVRED.

A SQVARE piece of land hath foure sides, or foure corners, whether they differ in widenesse or not, and it is either just squaire: That is, when the breadth is equall to the length, as is the first figure here following, or it is a long squaire as are the most parte of our Rigges of land, that is of a greater length nor breadth, conforme to the next figure following.

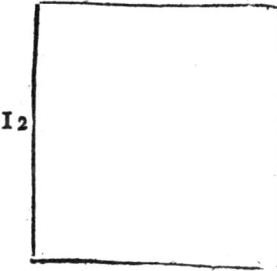

C The

Of Measuring of land.

```
            30
┌─────────────────────┐
│                     │ 6
└─────────────────────┘
```

The firſt figure is vppon all ſides equall, that is 12. Fallis on every ſide. To find the quantitie thereof by Arithmetik you muſt multiply the length by the breadth, which is 12 by 12 : Thereof ariſeth 144 Fallis, which you ſhall devide by 40 Fallis, becauſe 40 Fallis is a Roode, and you will finde that it extends to 3 Roodes and 24 Fallis: or otherwiſe to ſeeke the compt thereof in the Table, if you can not multiplie nor devide numbers : and you will finde in the Columne of twelue Fallis of breadth, againſt the number of twelue Fallis in length 3 Roodes and 24 Fallis as ſaid is. The ſecond figure, is a long ſquare, not equall on all ſides, yet equall in breadth at both the ends, and equall of length at both the ſides, the length thereof is 30 Fallis, and the breadth 6 Fallis : To know the compt thereof by the Table, you ſhall ſeeke the Columne of 6 Fallis in breadth: and you will finde againſt the number of 30 in length an Acre and twentie Fallis for the quantitie thereof.

HOW TO MEASVRE LAND THAT
is of an vnequall breadth, and to finde
the compt thereof.

THere is foure Rigges of land meaſured, and they are found to bee 40 Fallis in length : and becauſe

they

Of Measuring of land.

they are not of equall breadthe, wee measure the breadth thereof at two or three sundrie partes, as the fashion of the ground requires. The broadest part is found to bee ten Fallis in breadth, the narrowest part to bee but six Fallis, and the middle part eight Fallis of breadth. These three breadthes beeing cast together they extend to 24 Fallis, whereof the thrid part is 8 Fallis, which is the just breadth. Now to finde the quantitie thereof in the Table, seeke the number eight Fallis, which is the breadth vppon the head of the said Table, and in the same Columne against the number of 40 which is the length, you will finde two Acres for the quantitie of the said foure Rigges.

When you are to measure any croft land or Burrow Rigges, and can not espie by your eye, any difference in the breadth thereof, yet for trying of the trueth, you shall trie the breadth thereof the oftener, at everie 10 Fallis of the length at the least, and write everie one of them particularly, and suppose that you haue taken the breadth at 6 sundrie times, you shall add them all in one summe, and then devide that summe in 6 partes, and take that sixt part for your breadth, and with that breadth and the just length resort to the Table.

TO MEASVRE LAND THAT IS vnequall both in length and breadth.

A Piece of land being vnequall at all partes, is measured at both the sides, and at both the endes, the length of the longest side is 16, and the shortest side is 10, the breadth at the broadest end is 4, and at the narrow end 2. Now add the two lengths together, as 16 and 10 makes 26. Take the halfe thereof which is 13, for the length: and add the two breadths together

together, as 4. and 2. makes 6 Take the halfe thereof which is 3. for the breadth, and then with 3. of breadth and 13 of length, reforte to the Table: in the Columne of 3 Fallis againſt the length 13. and you will finde 39 Fallis, for the quantitie of this piece of land here following.

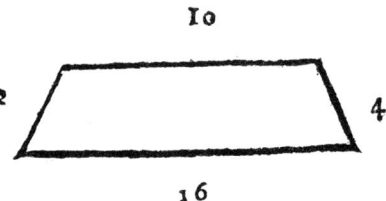

OF TRIANGLES, AND THE WAY
how to meaſure them.

A piece of land is called a Triangle, that is imagined to haue three ſides and three corners: whether the ſides bee equall or otherwiſe. There is no piece of land, but it may bee caſten in Triangles, and ſo moſt truelie meaſured. And becauſe it is requiſite, that in the meaſuring of all Triangles, firſt to finde a right hanging or deſcending line in everie Triangle: by the helpe of the which Line all landes of Triangle faſhion are brought to bee meaſured, and therefore the manner is here ſet downe.

HOW

Of measuring of land.

HOW THE RIGHT DESCENDING Line, is drawne in Triangles.

HE said Lyne is ever drawne, or imagined to come downe Square-wayes, from any corner of the Triangle to some of the sides thereof, as the descending Lyne in this figure following, betwixt a. and b. cutteth this Triangle, in the Lyne, c. d. Square-wayes in the point. b. and not as the other Lyne a. and e. doeth,

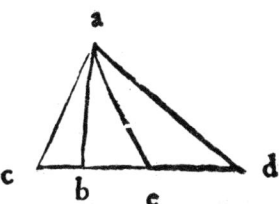

After that the said descending Lyne is drawne, then to measure any Triangle, you shall first measure the lengthe of the said Lyne, and then measure the lengthe of that side of the Triangle, that the said Lyne cutteth Square wayes. This done, Take the halfe of the measure of any of the saids Lynes, with the whole measure or length of the other Lyne, and with them as with the length and breadth resort to the Table, in manner following,

EXAMPLE.

You shall imagine this Triangle following, that it is marked vpon the corners with a. b. c. d. to bee a peece of land whereof you desire to know the just quantitie. It is found that the descending Lyne, that is

brought

Of Meaſuring of land.

brought from the corner a. to the ſide b. c. and meetteth Square at the point d. to be 24. Falles in lengthe and the ſide betweene b. c. to bee 40. Falles in length.

Now take the halfe of the ſaid deſcending Lyne, which is 12 Falles, and the length of the ſaid Lyne b. c. which is 40. Falles, and reſort to the Table with 12. in breadth and 40. in length, and in the Columne of 12. Falles of breadth, you will finde againſt 40 of lēgth, 3 Acres for the quātitie of this triangle folowing.

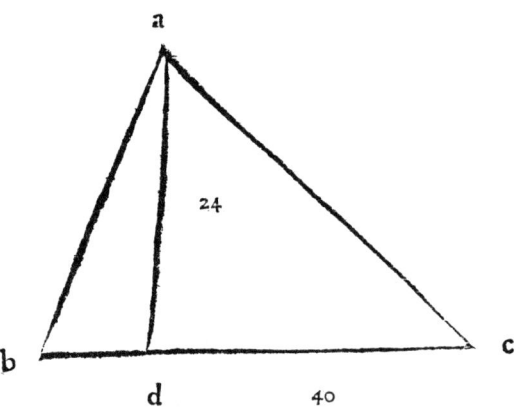

It is not needfull to the common ſort of people, to trouble themſelves to know any further concerning the Meaſuring of Land, but onlie to vnderſtand theſe former bypaſt examples which may ſerve them for inſtruction to know how to meaſure all their Lands, and to finde the compt, what they containe in Acres, and other ſmaller quantities, But theſe other examples following, as of vnequall peeces of Land, of Rounds,

Of measuring of land.   23

of *Triangle squares*, of *Woodes*, *Hilles*, and *Valleyes*, with the examples of small measures both in length and the breadth is set downe to give contentment, to such as are curious, to be resolved how they shall measure, and finde the quantitie of such fashions of Land, in case they shall haue to do therewith.

THE RVLE HOW TO MEAsure vnequall peeces of Land.

WHen any peece of Land happeneth to be lying in such forme, that it hath many vnequall points, and corners. Then because it is neither Triangle, nor Square, vntil it bee divided, and casten in three, or foure partes, as it will require. There is heere set downe three imagined peeces of Land, to bee reduced in Triangles, or Squares, and then measured by the ordr of the rules before specified.

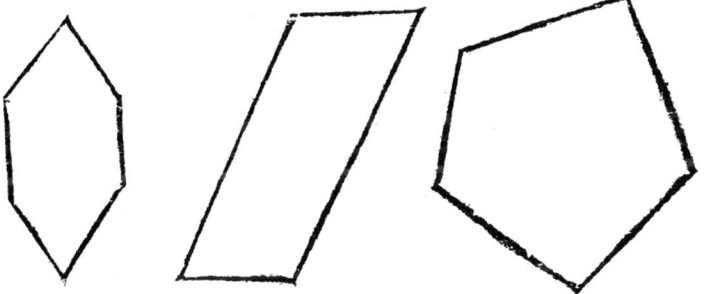

HEERE

Of meaſuring of Land.

HEERE FOLLOWETH EX-
amples how the foreſaids figurs of vn-
equall forme ſhould bee divided, &
reduced, in manner foreſaid.

YOV may perceave that the firſt figure is caſten into a Square, and two Triangles. The ſecond figure is divided into two Triangles, and the third figure in three Triangles. Now after that you haue ſeparated and caſten the firſt figure in manner foreſaide, then you are to knowe the manner how to meaſure it, a n d thereafter to finde the quantitie thereof. The manner how to meaſure it, is firſt, to mett and meaſure the Square peece, and thereafter the two Triangles.

Now I ſuppone that the Square peece is found to be on everie ſide twentie Falles, which is both lengthe, and breadth. To know the quantitie thereof, you will find in the Columne of twentie Falles in breadthe, againſt the number of twentie Falles in lengthe two Acres, and two Roode, for the quantitie of the Square peece, and to meaſure the two Triangles at the ends of the ſaide Square, the deſcending Lyne of the vpper Triangle, is ſuppoſed to bee twelue Falles. The halfe whereof is ſixe Falles for the breadthe, and the nether end of the ſaid Triangle to be twentie Falles. Reſort to the Table with ſixe Falles in breadthe, and twentie in lengthe, and you will finde three Roodes, for the quantitie of the ſaid vpper Triangle. Now to know the quantitie of the nether Triangle, the deſcending Lyne whereof, is ſuppoſed to bee foureteene Falles. The halfe whereof, is ſeven Falles for the breadthe, and the ſide of the Triangle to bee twentie Falles. Reſort to the Table with ſeven Falles in breadthe, and

Of Measuring of land.   25

twentie in length, and you will finde three Roodes, and 20. Fallis for the quantitie of the nether Triangle. Now cast all these three summes into one *viz.* the quantitie of the square piece is two Acres two Roodes, with the quantitie of the vpper Triangle, which is 3 Roodes, and the quantitie of the nether Triangle, 3 Roodes 20. Fallis: they all extend to 4 Acres and 20 Fallis, which is the quantitie of the said first figure, here devided in this forme following.

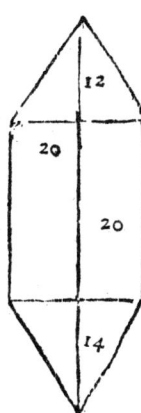

ALSO you see the second figure to bee devided in 2 parts of Triangle land, the descending line of the vpper Triangle, is supposed to bee eight Fallis: the halfe whereof, to wit, foure is the just breadth of the vpper Triangle: and the line that seperates the two Triangles, to be 40 Fallis for the length of the said Triangle: So foure Fallis for the breadth, with 40 for the

D   length,

length, being brought to the Table, maketh the first Triangle to bee a just Acre of land. Now suppose the descending line of the nether Triangle to be 10 Fallis in breadth, the halfe whereof is fiue, and the said line of separation being 40 for the length, which being sought in the Table, will be an Acre and a Roode for the quantitie of the nether Triangle: So the quantitie of both is two Acres and a Roode for the quantitie of this figure.

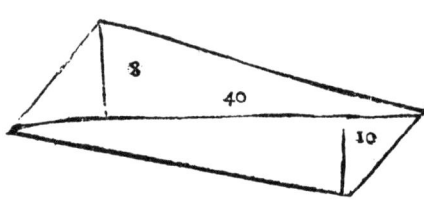

Also you see the thrid figure here following, is devided by the two lines of separation into three Triangles, which must be measured after the same manner, in the manner of Triangles, and compted by the Table with length and breadth as said is: And thus much for avoyding of tediousnesse.

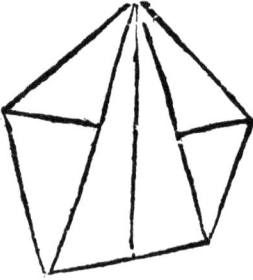

THE

THE RVLE FOR MEASVRING
of Circles, or Round peeces of Land.

A Round peece of Land is without corner, or square, and is called a Circle. The compaſſe thereof, is called the Circumference. The middle point is the Center. The Lyne going thorow the Center, or middeſt of the Circle, touching it on both ſides, is called the Diameter: the half whereof, is called Semidiameter, and a peece of a Circle, is called an Arke. For meaſuring of all rounds, take the halfe of the Diameter for the breadth, and the halfe of the circumference for the length, and therewith reſort to the Table to finde the quantitie. Example, Imagine this preſent round figure to bee a peece of Land. The Circumference whereof to bee an hundreth, and twentie Falles, and the Diameter to be 40 Falles: take the half of the Diameter, which is 20 Falles for the breadth, and the halfe of 120 Falles, which is 60. Falles for the lengthe, reſort to the Table therewith, and you will find 7 Acres, and 2 Roods, for the quantitie of this Circle.

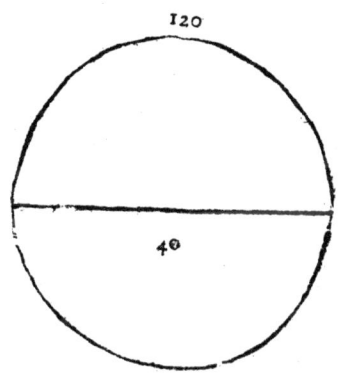

28　Of Measuring of Land.

AS for measuring of halfe roundes, you shall enter the Table with the halfe of the Circumference thereof for the length, and with the halfe Diameter for the breadth. Example, the length of the halfe Diameter of this halfe Circle, is 20 Falles, and the halfe of the Circumferenc is 30 Falles, which being brought to the Table to the Columne of 20 Falles in breadth, you will finde right against the number 30 of length, three Acres, and 3 Roods, for the quantitie of this half Circle.

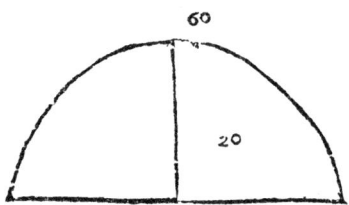

THE RVLE OF MEASVRING
Triangled Squares.

SOme peeces of land may fall out as these two figures following, and such like. And albeit they may divided and casten in Triangles, and so by the rule of Triangles measured, yet they haue their proper rule, and measuring as followeth. You shall joyne both the measures of the endes in one summe, and take the halfe of that number for the bredth, therafer measure the length of the peece, as you see the Lyne drawne through the middest heere. Example, Suppose the end of the litle peece to bee 8 Falles, and the nether end 12 Falles, they being joyned and added together are 20. The

halfe

Of Measuring of Land.

halfe whereof is 10, for the breadth, and the length of the middle Lyne, is 30 Falles. When you seeke the Table in the Columne of 10 Falles of breadth, you will find against the number 30. 1. Acre, and 3 Roodes: and 20 Falles for the quantitie of this least peece, and in like manner, you shall measure the other figure also.

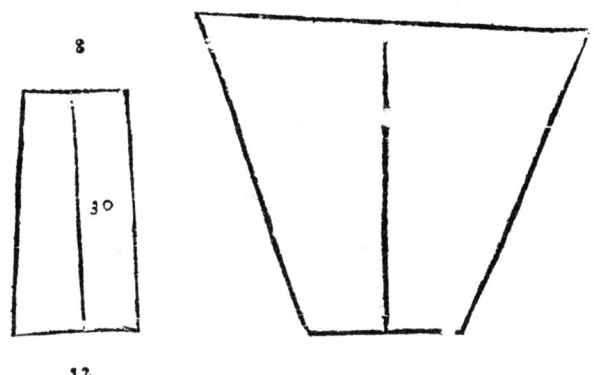

THE RVLE FOR MEASVRING OF *WOODS, MOSSES, AND LOCHES,* or other peeces of Land, which is deformed, and vnequall in all sides.

OR measuring such peeces of Land, as are evill fashioned, and cannot be measured, except it be casten in many Squares, and Triangles: then to save labour where Land is watrie, or can not bee seene for standing Wood, and such other impediments. For measuring thereof, or such other peeces of Land, as this present figure is, it shall bee best to adde

and

and joyne to the said peece of Land, so many portions at the deformed parts, as will make it square: or otherwise as you shall see this vnequall figure to bee heere following casten in a square.

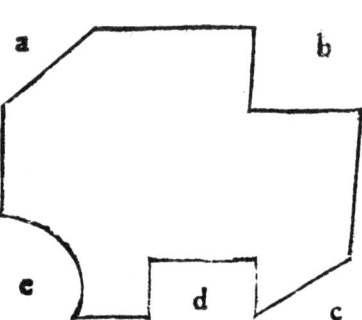

AS there is augmented in the part marked a, five Falles, and in the part marked b. three Falles, in c. sixe Falles, in d. three Falles, and in e. two Falles, all which peeces heere beeing augmented, and put in one summe are 19 Falles. Now suppone that in measuring the whole square, you finde the length to be 67 Falles, and the breadth 17: Then to know what number of Acres it containeth, you shall seeke the number of 17 Falles in the head of the Table. But because you haue not 67 Falles of length in one number, you shall take 60 of length, and then 7 of length both in that same Columne, and against the number of 60 you will finde 6 Acres, 1 Rood, and 20 Falles, and against the number of 7, you will find 2 Roodes, and 39 Falles, these being casten together will make 7 Acres, and

Of meafuring of land.

19 Falles, deduce the 19 Falles, that the faid peeces of augmented Land extens to, and there will reft 7 Acres for the quantitie of the faid peece of vnequall Land.

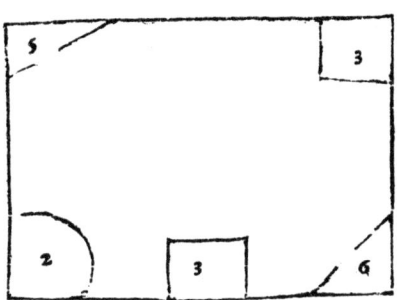

THE RVLE FOR MEASVRING
of mountaine Land, and of Valley ground. And firft of Hilles.

YOV fhall firft meafure the circuite of the bafe parte, or foote of the Hill, or Mountaine : Then meafure the toppe thereof, and adde them both together, fo mnft you doe with the afcenfe or fides of the Hill, that is to fay, the going vp from the foot of the Hill, to the top thereof, and put the meafure of the fhorteft and longeft together, and take the half of the faid afcenfe for the breadth, and the halfe

halfe of the circuit, or compasse of the foot and toppe of the Hill, for the length, As for example.

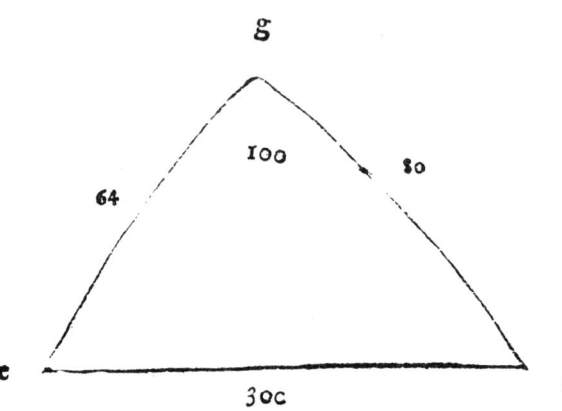

SVppone this figure marked e. f. g. to bee the mountain or Hill, the compasse of the foot thereof, is fund to be three hundreth Falles, the toppe thereof to bee 100 Falles, which are together 400 Falles for the length. Next the ascense betweene e and g. to bee 64 Falles, and the other ascense from f. to g. to bee 80 Falles. They beeing added together maketh 144 Falles, the halfe whereof, is 72 Falles for the iust breadth with these two summes we enter the Table, and becaufe there is no page, nor Columne of 72 Falles of breadth together, therefore you shall take the Columnes of 70. Falles, of breadth, and of 2 Falles, and in the Columne of 70 Falles against the length 200 Falles you will find 87 Acres, and 2 Roodes, and in the page, or Columne of 2 Falles of breadth, against the said number of 200 Falles

Of Measuring of Land. 33

Falles, you will finde 2 Acres, and 2 Roodes, which being added to the 87 Acres, and 2 Roodes, the whole is 90 Acres for the quantitie of this figured Hill.

THE RVLE FOR MEASVRING OF VALLEYES.

AS in the mountaine, or Hill, you measured the circuit or compasse of the foot of the Hill, so, heere contrariwise you shal met round about the circuit or compasse of the heigth of the Valley: And as you measured the toppe of the mountaine, So shall you measure the bottome, or depth of the Valley, In like manner as you measured the ascense or going vp from the foot of the Hill, to the toppe thereof: So should you measure heere the descense, or going downe to the bottome of the Valley. Example is heere figured of a peece of Land of an vnequall Valley, that it may bee the rule for others. Firſt, take the circuite of the height, which I suppose to bee 156 in the compasse about the top of the Valley, and the depth or bottome of the Valley to bee 24 Falles, adde them together they wil make 180. Falles, the halfe whereof is, 90. Falles for the breadth: then measure the descense or going downe of both the sides to the bottome, the one side wherof is, 152 Falles, the other side 188 they being added together are 340 Fals the half wherof is 170 Fals for the legth thē with 90 Falles of breadth, and 170 Falles of length, seeke the Table in the Columne of 90 Falles in breadth, you haue not the full number of 170 Falles of length in one summe: you shall firſt take 100 and next 70 againſt the number of 100, you will finde 56 Acres, and 1 Rood,

E

Of Measuring of land.

Rood, and against the number of 70, you will find 39 Acres, one Rood, and 20 Falles: adde these together they make 95 Acres, 2 Roodes, and 20 Falles, which is the quantitie of the said Valley.

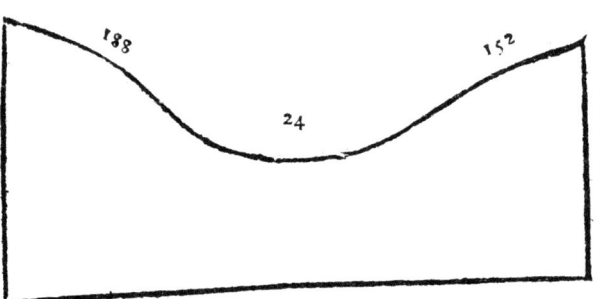

CONCERNING SMALL MEASVRES of Ells, when they shall happen out, in *measuring of land.*

BECAVSE oftentimes small measures, as the measure of 1. 2. 3. 4. 5. Ells of an halfe Ell, and quarter Ell will fall out often-times, to bee in the length and not in the breadth, and sometimes in the breadth and not in the length, and sometimes both in breadth and length as is said of before. The said small measures hath everie one of them their Columnes in the beginning of the Table: And their compt is sought and found out as other measures are: onely remembring that the small measures that falls out in the breadth, must be reckoned by them of the length, but the odd measures that happeneth in the length, must contrarie-wise bee reckoned with the breadth.

EXAM-

EXAMPLE OF SMALL MEASVRES in the breadth.

A PIECE of land is founde to bee 60. Fallis in length, and 10 Fallis and an Ell in breadth. Now to finde the quantitie thereof by the Table: you shall first seeke the Columne of the breadth of 10 Fallis, and then the Columne of the breadth of an Ell: and in the Columne of 10 Fallis of breadth, you will finde against the number 60 of length, 3 Acres and 3 Roodes: And in the Columne of an Ell in breadth, against the said number of 60, you will finde 10 Fallis. These being added together, they will extend to 3 Acres 3 Roodes and 10 Fallis, for the quantitie of that parsell of land as said is.

EXAMPLE OF SMALL MEASVRES in the length.

WHEN small measures shall happen to fall out in the length of any land, and not in the breadth, then you must remember to recken that with the bredth as said is. As for example, A piece of land is 80 Fallis fiue Ells, and a halfe Ell in length, and eight Fallis in breadth: To finde the true quantitie thereof, you shall seeke the breadth of 8 Fallis, and there right against the number of 80. you will finde 4 Acres: Then you shall seeke the Columne of the small measures in the length, which is fiue Ells and a halfe: And in the Columne of fiue Ells, which you must recken as with the breadth as said is, against the number of 8, you will finde 6 Fallis 4 Ells: And in the Columne of an halfe Ell, you will finde against the number of 8. 4. Ells. These beeing

E 2 added

Of measuring of Land.

added together, they will make 4. Acres 7. Fallis. 2. Ells for the quantitie of that piece, as said is.

Before any examples bee set forth, anent small measures both in the length, and in the breadth: it is needefull to set downe a small Table, to resolve the compt when small measures, should be multiplyed and compted with small measures, which the former Table hath not.

A DESCRIPTION OF THE TABLE of small measures.

THIS Table is set downe, to resolve such men as will bee precise to knowe the extremitie of the compt of measured land: it containeth nine Columnes. The first Columne vpon the left hand, containeth everie quantitie of small measures, descending downe from the head to the foote of the Table, as of one quarter, which is a quarter of an Ell, Next of halfe, which is halfe an Ell: Then of three quarters, which is three quarters of an Ell, and of 1. 2. 3. 4. and 5. Ells, which are the whole small measures, that can fall out in the measuring of land. The head of the said Table hath the same measures beginning at the left hand, going forth orderlie to the right hand. Nowe for trying thereof by example: In case you haue 3 Ells of small measures in the breadth, to bee compted with 4 Ells in the length, you shall seeke the number 3 in the side of the Table, and goe right forth from it to the right hand, vntrll you come vnder the number 4, standing vpon the head of the Table: And in that Square where they meet, you will finde 2 Ells, for the quantitie they extend to:

Or

Of measuring of land. 37

Or otherwise seeke the number 3. vppon the head of the Table, and come right downe, vntill you come against 4, standing vppon the left side of the Table, and in that Square where they meete, you will finde the same answer of 2 Ells: Or if you haue three quarters of an Ell, to bee compted with a halfe Ell, you will finde in the Square where they meete, the sixteene part of an Ell, which wee call a naile for the quantitie thereof. The most part of the quantities that are in the said Table, consistes in fractions or broken numbers, which may bee easilie read and vnderstood, if you will consider them in this manner, you will finde two 8 partes: that is, if an Ell were devided in 8 partes, it is 2 of these 8 partes. Also you will finde five 24 partes: that is, if an Ell were devided in 24 partes, it is 5 of these 24 partes: And thus much for vnderstanding of these & all broken numbers.

Now becaufe it may bee said, that this small Table is not rightly set downe, becaufe that 4 Ells being multiplyed by 3 Ells should yeeld 12 Ells, and this small Table produces but 2 Ells. I answer that this small Table produces but the sixt part of the full number, becaufe it is made onely to resolve the quantitie that ariseth vpon small measures, when they are multiplyed one by the other, not being in their owne strength, but standing after greater measures as partes thereof: As 8 Fallis 4 Ells of length to bee multiplyed by 4 Fallis 3 Ells of breadth: There the 4 Ells standing after the 8 Fallis is but one part of a Fall of measure: To wit, the two thrid part of a Fall, and the 3 Ells is but the halfe of a Fall. Now to multiplie two thrids of a Fall, with the halfe of a Fall, they will produce but one thrid part of a Fall, which is 2 Ells as is set downe in this small Table following: And thus much for resolving of that doubt.

Of Meaſuring of land.

	1/4 Eln	1/2 Eln	3/4	1 Eln	2	3	4	5 Eln
1/4	1/90	1/48	1/32	1/24	1/12	1/8	1/6	5/24
1/2	1/48	1/24	1/16	1/12	1/6	1/4	1/3	5/12
3/4	1/32	1/16	3/32	1/8	1/4	3/8	1/2	5/8
1	1/24	1/12	1/8	1/6	1/3	1/2	2/3	5/6
2	1/12	1/6	1/4	1/3	3/3	1 Eln	1 1/3	1 2/3
3	1/8	1/4	3/8	1/2	1 Eln	1 1/2	2 Eln	2 1/2
4	1/6	1/3	1/2	2/3	1 1/3	2	2 2/3	3 1/3
5	5/24	5/12	5/8	5/6	1 2/3	2 1/2	3 1/3	4 1/6

EXAMPLE OF SMALL MEASVRES both in the length, and in the breadth.

THESE forme of examples will not fall out, except that men will bee preciſe and extreame in the meaſuring of their land: yet it is needfull to reſolve ſuch preciſe men by an Example, leaſt they thinke that ſuch queſtions can not be reſolved by Table.

Of measurnig Land. 39

And albeit it will seeme to be obscure to vnderstand, and difficile to finde out the trueth of such a compt: if you will take heede and seeke it in manner following, you will find it more easily, truely, and perfitely compted by the Tables, than it can be done by tongue, as the Landmetters are accustomed to do. The maner then to find it is, First you shall cast vp in the great Table the particular columnes both great & smal of the whole bredths and seeke into everie one of them the compt of the greatest length. Secondly, you shall cast vp in the great Table the columnes of the small measures that shall happen to fall out in any land, and seek into every one of them the compt of the greatest breadth. Thirdly, resort o this small Table, and seeke after the maner set downe in the description thereof: in every columne of the small measures of the breadth, the compt of every small measure of the length, and adde all these summes together.

EXAMPLE.

A peece of Land containeth 80 Falles, 4 Ells, and a half Ell of length, and 8 Falls, 3 Ells, and 3 quarters of an Ell in breadth: To know what the extremity that compt will extend to, you shall seek the Tables in maner foresaid, that is, first to cast vp in the great Table the columne of 8 Falles, the columne of 3 Ells, and the columne of 3 quarters, which are the columnes of the whole breadths in the said columne of 8 Falles, seek the length 80, you will find against the the said number standing 4 Acres, in the columne of 3 Ells, against 80, you will find 1 Roode, and in the columne of 3 quarters of an Ell, against 80 you will find 10 Falles: These being added together will extend to 4 Acres, 1 Rood and 10 Falles. Secondly, cast vp in the said great Table, the columne of 4 Ells, and the columne of an half Ell, which are the columnes of the small mea-

measures of the length, in the columne of 4 Ells, seek the greatest breath, which is 80, and there against the saide number, you will find 5 Falles, 2 Ells, and in the columne of an halfe Ell, against the said number of 8, you will find 4 Ells: these two being added together will make 6 Falles. Thirdly, resort to the small Table, and seek according as it is set downe in the description thereof the columne of 3 Elles, and in the columne of 3 quarters of an Ell, which are the small measures of the breadth, in the columne of 3 Ells seek 4 Ells, and an halfe Ell against 4 Ells, you will find 2 Ells, and against the halfe Ell you will find a quarter of an Ell, then in the said columne of 3 quarters against 4 you will find an half Ell, and against an halfe Ell, you will find a 16 part of an Ell, these being both added together will extend to a very smal quantitie, which is, 3 quarters and 16 part Ell. The whole 3 compts being added together will extend to 4 Acres, 1 Roode, 16 Failes, 2 Ells, 3 quarters of an Ell and a Naill. And this is the just compt thereof, which may serue for all others of the like forme.

OBSERVATION.

This Table following will resolve you of sundry questions, as if a peece of land were 40 Falles in bredth & length, if you desire it to be a park right square, & to know what Acres it will contein you shall seek the bredlth 40, & against 40 of length, you will find 10 Acres for the quantitie therof. If you haue a peece land 24 Falles of bredth, to contein 9 Acres, to know what length to take to make it 9 Acres or thereby, seek the columne of 24 Falles, & come down that columne in the broad roome till you find 9 Acres, & you finde 9 Acres standing against 60, so you must take 60 Falls of length to 24 Falles of bredth to make the park 9 Acres of land, and so furth vse it in any number.

You are to be remembred that Glasen-wrigths worke, may be compted with the Table made for Sclaters, and Masons, compting Footes, in stead of Elles.

THE

The Compt of meaſured Land. 41

Quarter Elne.			Half Elne.			3 Quar. Elne.			1 Elne.	
	Quat.			El. Quar.			El. Qu.			El.
1		1	1		2.	1		3.	1	1
2	El.	2	2	1.	0.	2	1.	2.	2	2
3		3	3	1.	2.	3	2.	1.	3	3
4	1. 0.		4	2.	0.	4	3.	0.	4	4
5	1. 1.		5	2.	2.	5	Falles 3.	3.	5	5
6	1. 2.		6	3.	0.	6	4.	2.	6	Falles 0
7	1. 3.		7	3.	2.	7	5.	1.	7	1. 1
8	2. 0.		8	4.	0.	8	1. 0.	0.	8	1. 2
9	2. 1.		9	4.	2.	9	1. 0.	3.	9	1. 3
10	2. 2.		10	Falles 5.	0.	10	1. 1.	2.	10	1. 4
11	2. 3.		11	5.	2.	11	1. 2.	1.	11	1. 5
12	3. 0.		12	1. 0.	0.	12	1. 3.	0.	12	2. 0
13	3. 1.		13	1. 0.	2.	13	1. 3.	3.	13	2. 1
14	3. 2.		14	1. 1.	0.	14	1. 4.	2.	14	2. 2
15	3. 3.		15	1. 1.	2.	15	1. 5.	1.	15	2. 3
16	4. 0.		16	1. 2.	0.	16	2. 0.	0.	16	2. 4
17	4. 1.		17	1. 2.	2.	17	2. 0.	3	17	2. 5
18	4. 2.		18	1. 3.	0.	18	2. 1.	2	18	3. 0
19	4. 3.		19	1. 3.	2.	19	2. 2.	1	19	3. 1
20	5. 0.		20	1. 4.	0.	20	2. 3.	0	20	3. 2
21	5. 1.		21	1. 4.	2.	21	2. 3.	3	21	3. 3
22	5. 2.		22	1. 5.	0.	22	2. 4.	2	22	3. 4
23	5. 3.		23	1. 5.	2.	23	2. 5.	1	23	3. 5
24	Falles 6.	0	24	2. 0.	0.	24	3. 0.	0	24	4. 0
25	6.	1.	25	2. 0.	2.	25	3. 0.	3	25	4. 1
30	1. 1. 2.		30	2. 3.	0.	30	3. 4.	2	30	5. 0
40	1. 4. 0.		40	3. 2.	0.	40	5. 0.	0	40	6. 4
50	2. 0. 2.		50	4. 1.	0.	50	6. 1.	2	50	8. 2
60	2. 3. 0.		60	5. 0.	0.	60	7. 3.	0	60	10. 0
70	2. 5. 2.		70	5. 5.	0.	70	8. 4.	2	70	11. 4
80	3. 2. 0.		80	6. 4.	0.	80	10. 0.	0	80	13. 2
90	3. 4. 2.		90	7. 3.	0.	90	11. 1.	2	90	15. 0
100	4. 1. 0.		100	8. 2.	0.	100	12. 3.	0	100	16. 4
200	8. 2. 0.		200	16. 4.	0.	200	25. 0.	0	200	33. 2

F

The compt of measured Land.

	Two Elnes.			Three Elnes.			Foure Elnes.			Five Elnes.	
Fal.	Fal.	El.	Fal.	Fal.	El.	Fal.	Fal.	El.	Fal.	Fal.	El.
1		2	1		3	1		4	1		9
2		4	2	1.	0	2	1.	2	2	1.	4
3	1.	0	3	1.	3	3	2.	0	3	2.	3
4	1.	2	4	2.	0	4	2.	4	4	3.	2
5	1.	4	5	2.	3	5	3.	2	5	4.	1
6	2.	0	6	3.	0	6	4.	0	6	5.	0
7	2.	2	7	3.	3	7	4.	4	7	5.	5
8	2.	4	8	4.	0	8	5.	2	8	6.	4
9	3.	0	9	4.	3	9	6.	0	9	7.	3
10	3.	2	10	5.	0	10	6.	4	10	8.	2
11	3.	4	11	5.	3	11	7.	2	11	9.	1
12	4.	0	12	6.	0	12	8.	0	12	10.	0
13	4.	2	13	6.	3	13	8.	4	13	10.	5
14	4.	4	14	7.	0	14	9.	2	14	11.	4
15	5.	0	15	7.	3	15	10.	0	15	12.	3
16	5.	2	16	8.	0	16	10.	4	16	13.	2
17	5.	4	17	8.	3	17	11.	2	17	14.	1
18	6.	0	18	9.	0	18	12.	0	18	15.	0
19	6.	2	19	9.	3	19	12.	4	19	15.	5
20	6.	4	20	10.	0	20	13.	2	20	16.	4
21	7.	0	21	10.	3	21	14.	0	21	17.	3
22	7.	2	22	11.	0	22	14.	4	22	18.	2
23	7.	4	23	11.	3	23	15.	2	23	19.	1
24	8.	0	24	12.	0	24	16.	0	24	20.	0
25	8.	2	25	12.	3	25	16.	4	25	20.	5
30	10.	0	30	15.	0	30	20.	0	30	25.	0
40	13.	2	40	20.	0	40	26.	4	40	33.	2
50	16.	4	50	25.	0	50	33.	2	50	1. 1.	4
60	20.	0	60	30.	0	60	1. 0.	0	60	1. 10.	0
70	23.	2	70	35.	0	70	1. 6.	4	70	1. 18.	2
80	26.	4	80	1. 0.	0	80	1. 13.	2	80	1. 26.	4
90	30.	0	90	1. 5.	0	90	1. 20	0	90	1. 35.	0
100	33.	2	100	1. 10.	0	100	1. 26	4	100	2. 3.	2
200	1. 26.	4	200	2. 20.	0	200	3. 13.	2	200	4. 6.	4

The compt of meafured Land.   43

1. Fall.		2. Falles.		3. Falles.		4. Falles.	
Fall.	Falles		Fal.		Falles.		Fal.
1	1	1	2.	1	3.	1	4
2	2	2	4.	2	6.	2	8
3	3	3	6.	3	9.	3	12
4	4.	4	8.	4	12.	4	16
5	5.	5	10.	5	15.	5	20
6	6.	6	12.	6	18.	6	24
7	7.	7	14.	7	21.-	7	28
8	8.	8	16.	8	24	8	32
9	9.	9	18.	9	27.	9	36
10	10.	10	20.	10	30.	10	Roodes 1.	0
11	11.	11	22.	11	33	11	1.	4
12	12	12	24.	12	36	12	1.	8
13	13	13	26.	13	39.	13	1.	12
14	14	14	28.	14	Roodes 1.	2	14	1.	16
15	15.	15	30.	15	1.	5	15	1.	20
16	16.	16	32	16	1.	8	16	1.	24
17	17.	17	34.	17	1	11	17	1.	28
18	18.	18	36.	18	1.	14	18	1.	32
19	19.	19	38	19	1	17	19	1.	36
20	20.	20	Roodes 1.	0	20	1	20	20	2.	0
21	21.	21	1.	2.	21	1.	23	21	2.	4
22	22.	22	1.	4.	22	Acres 1.	26	22	2.	8
23	Roodes. 23.	23	1.	6.	23	1.	29	23	2,	12
24	24.	24	1.	8.	24	1.	32	24	2.	16
25	25.	25	1.	10.	25	1.	35	25	Acres 2.	20
30	30.	30	1.	20.	30	2.	10	30	3.	0
40	1.	0.	40	Acres 2.	0.	40	3.	0	40	1.	0.	0
50	1.	10	50	2.	20.	50	3.	30	50	1.	1.	0
60	1.	20	60	3.	0.	60	1	20	0	60	1.	2.	0
70	1.	30	70	3.	20.	70	1	1.	10	70	1.	3.	0
80	2.	0.	80	1.	0.	0.	80	1	2.	0	80	2.	0.	0
90	2.	10.	90	1.	0.	20.	90	1	2.	30	90	2,	1.	0
100	2.	20	100	1.	1.	0.	100	1	3.	20	100	2.	2.	0
200	5.	0	200	2.	2.	0.	200	3	3.	0	200	5	0.	0

F 2

The compt of measured Land.

Fal.	5. Falles				6. Falles				7. Falles				8. Falles		
	Acres	Roodes	Fal.	Fal.	Acres	Roodes	Fal.	Fal.	Acres	Roodes	Fal.	Fal.	Acres	Roodes	Fal.
1			5	1			6	1			7	1			8
2			10	2			12	2			14	2			16
3			15	3			18	3			21	3			24
4			20	4			24	4			28	4			32
5			25	5			30	5			35	5		1.	0
6			30	6			36	6		1.	2	6		1.	8
7			35	7		1.	2	7		1.	9	7		1.	16
8		1.	0	8		1.	8	8		1.	16	8		1.	24
9		1.	5	9		1.	14	9		1.	23	9		1.	32
10		1.	10	10		1.	20	10		1.	30	10		2.	0
11		1.	15	11		1.	26	11		1.	37	11		2.	8
12		1.	20	12		1.	32	12		2.	4	12		2.	16
13		1.	25	13		1.	38	13		2.	11	13		2.	24
14		1.	30	14		2.	4	14		2.	18	14		2.	32
15		1.	35	15		2.	10	15		2.	25	15		3.	0
16		2.	0	16		2.	16	16		2.	32	16		3.	8
17		2.	5	17		2.	22	17		2.	39	17		3.	16
18		2.	10	18		2.	28	18		3.	6	18		3.	24
19		2.	15	19		2.	34	19		3.	13	19		3.	32
20		2.	20	20		3.	0	20		3.	20	20	1.	0.	0
21		2.	25	21		3.	6	21		3.	27	21	1.	0.	8
22		2.	30	22		3.	12	22		3.	34	22	1.	0.	16
23		2.	35	23		3.	18	23	1.	0.	1	23	1.	0.	24
24		3.	0	24		3.	24	24	1.	0.	8	24	1.	0.	32
25		3.	5	25		3.	30	25	1.	0.	15	25	1.	1.	0
30		3.	30	30	1.	0.	20	30	1.	1.	10	30	1.	2.	0
40	1.	1.	0	40	1.	2.	0	40	1.	3.	0	40	2.	0.	0
50	1.	2.	10	50	1.	3.	20	50	2.	0.	30	50	2.	2.	0
60	1.	3.	20	60	2.	1.	0	60	2.	2.	20	60	3.	0.	0
70	2.	0.	30	70	2.	2.	20	70	3.	0.	10	70	3.	2.	0
80	2.	2.	0	80	3.	0.	0	80	3.	2.	0	80	4.	0.	0
90	2.	3.	10	90	3.	1.	20	90	3.	3.	30	90	4.	2.	0
100	3.	0.	0	100	3.	3.	0	100	4.	1.	20	100	5.	0.	0
200	6.	1.	0	200	7.	2.	0	200	8.	3.	0	200	10.	0.	0

The compt of meafured Land. 45

Fal.	9. Falles.		Fal.	10. Falles.		Fal.	11. Falles.		Fal.	12. Falles.	
1	Rood	9	1	Rood	10	1	Rood	11	1	Rood	12
2		18	2		20	2		22	2		24
3		27	3		30	3		33	3		36
4		36	4	1.	0	4	1.	4	4	1.	8
5	1.	5	5	1.	10	5	1.	15	5	1	20
6	1.	14	6	1.	20	6	1.	26	6	1.	32
7	1.	23	7	1.	30	7	1.	37	7	2.	4
8	1.	32	8	2.	0	8	2.	8	8	2.	16
9	2.	1	9	2.	10	9	2.	19	9	2.	28
10	2.	10	10	2	20	10	2.	30	10	3.	0
11	2.	19	11	2.	30	11	3.	1	11	3.	12
12	2.	28	12	3.	0	12	3.	12	12	3.	24
13	2.	37	13	3.	10	13	3.	23	13	3.	36
14	3.	6	14	3.	20	14	3.	34	14	1. 0.	8
15	Acres 3.	15	15	Acres 3.	30	15	1. 0.	5	15	1. 0.	20
16	3.	24	16	1. 0.	0	16	1. 0.	16	16	1. 0.	32
17	3.	33	17	1. 0.	10	17	1. 0.	27	17	1. 1.	4
18	1. 0.	2	18	1. 0.	20	18	1. 0.	38	18	1. 1.	16
19	1. 0.	11	19	1. 0.	30	19	1. 1.	9	19	1. 1.	28
20	1. 0.	20	20	1. 1.	0	20	1. 1.	20	20	1. 2.	0
21	1. 0.	29	21	1. 1.	10	21	1. 1.	31	21	1. 2.	12
22	1. 0.	38	22	1. 1.	20	22	1. 2.	2	22	1. 2.	24
23	1. 1.	7	23	1. 1.	30	23	1. 2.	13	23	1. 2.	36
24	1. 1.	16	24	1. 2.	0	24	1. 2.	24	24	1. 3.	8
25	1. 1.	25	25	1. 2.	10	25	1. 2.	35	25	1. 3.	20
30	1. 2.	30	30	1. 3.	20	30	2. 0.	10	30	2. 1.	0
40	2. 1.	0	40	2. 2.	0	40	2. 3.	0	40	3. 0.	0
50	2. 3.	10	50	3. 0.	20	50	3. 1.	30	50	3. 3.	0
60	3. 1.	20	60	3. 3.	0	60	4. 0.	20	60	4. 2.	0
70	3. 3.	30	70	4. 1.	20	70	4. 3.	10	70	5. 1.	0
80	4. 2.	0	80	5. 0.	0	80	5. 2.	0	80	6. 0.	0
90	5. 0.	10	90	5. 2.	20	90	6. 0	30	90	6. 3.	0
100	5. 2.	20	100	6. 1.	0	100	6. 3.	20	100	7. 2.	0
200	11. 1.	0	200	12. 2.	0	200	13. 3.	0	200	15. 0.	0

The compt of measured Land.

Fal.	13. Falles.		Fal	14. Falles.		Fal.	15. Falles.		Fal.	16. Falles.		Fal.
	Rood	Fal		Rood	Fal.		Rood	Fal			Ro.	Fal.
1		13	1		14	1		15	1			16
2		26	2		28	2		30	2			32
3		39	3	1.	2	3	1.	5	3		1.	8
4	1.	12	4	1.	16	4	1.	20	4		1.	24
5	1.	25	5	1.	30	5	1.	35	5		2.	0
6	1.	38	6	2.	4	6	2.	10	6	Acres	2.	16
7	2.	11	7	2.	18	7	2.	25	7		2.	32
8	2.	24	8	2.	32	8	3.	0	8		3.	8
9	2.	37	9	3.	6	9	3.	15	9		3.	24
10	3.	10	10	3.	20	10	3.	30	10	1.	0.	0
11	3.	23	11	3.	34	11	1. 0.	5	11	1.	0.	16
12	3.	36	12	1. 0.	8	12	1. 0.	20	12	1.	0.	32
13	1. 0.	9	13	1. 0.	22	13	1. 0.	35	13	1.	1.	8
14	1. 0,	22	14	1. 0.	36	14	1. 1.	10	14	1.	1.	24
15	1. 0.	35	15	1. 1.	10	15	1. 1.	25	15	1.	2.	0
16	1. 1.	8	16	1. 1.	24	16	1. 2.	0	16	1.	2.	16
17	1. 1.	21	17	1. 1.	38	17	1. 2.	15	17	1.	2.	32
18	1. 1.	34	18	1. 2.	12	18	1. 2.	30	18	1.	3.	8
19	1. 2.	7	19	1. 2.	26	19	1. 3.	5	19	1.	3.	24
20	1. 2.	20	20	1. 3.	0	20	1. 3.	20	20	2.	0.	0
21	1. 2.	33	21	1. 3.	14	21	1. 3.	35	21	2.	0.	16
22	1. 3.	6	22	1. 3.	28	22	2. 0.	10	22	2.	0.	32
23	1. 3.	19	23	2. 0.	2	23	2. 0.	25	23	2.	1.	8
24	1. 3.	32	24	2. 0.	16	24	2. 1.	0	24	2.	1.	24
25	2. 0.	5	25	2. 0.	30	25	2. 1.	15	25	2.	2.	0
30	2. 1.	30	30	2. 2.	20	30	2. 3.	10	30	3.	0	0
40	3. 1.	0	40	3. 2.	0	40	3. 3.	0	40	4.	0.	0
50	4. 0,	10	50	4. 1.	20	50	4. 2.	30	50	5.	0.	0
60	4. 3.	20	60	5. 1.	0	60	5. 2.	20	60	6.	0.	0
70	5. 2.	30	70	6. 0.	20	70	6. 2.	10	70	7.	0.	0
80	6. 2.	0	80	7. 0.	0	80	7. 2.	0	80	8.	0.	0
90	7. 1.	10	90	7. 3	20	90	8. 1.	30	90	9.	0.	0
100	8. 0.	20	100	8. 3.	0	100	9. 1.	20	100	10.	0.	0
200	16. 1.	0	200	17. 2.	0	200	18. 3	0	200	20.	0.	0

The compt of measured Land. 47

	17. Falles.			18. Falles.			19. Falles.			20. Falles.	
Fal.	Rood	Fal	Fal.	Rood	Fal.	Fal.	Rood	Fal	Fal.	Ro.	Fal.
1		17	1		18	1		19	1		20
2		34	2		36	2		38	2	1.	0
3	1.	11	3	1.	14	3	1.	17	3	1.	20
4	1.	28	4	1.	32	4	1.	36	4	2.	0
5	2.	5	5	2.	10	5	2.	15	5	2.	20
6	2.	22	6	2.	28	6	2.	34	6	3.	0
7	2.	39	7	Acres 3.	6	7	Acres 3.	13	7	Acres 3.	20
8	Acres 3.	16	8	3.	24	8	3.	32	8	1. 0.	0
9	3.	33	9	1. 0.	2	9	1. 0.	11	9	1. 0.	20
10	1. 0.	10	10	1. 0.	20	10	1. 0.	30	10	1. 1.	0
11	1. 0.	27	11	1. 0.	38	11	1. 1.	9	11	1. 1.	20
12	1. 1.	4	12	1. 1.	16	12	1. 1.	28	12	1. 2.	0
13	1. 1.	21	13	1. 1.	34	13	1. 2.	7	13	1. 2.	20
14	1. 1.	38	14	1. 2.	12	14	1. 2.	26	14	1. 3.	0
15	1. 2.	15	15	1. 2.	30	15	1. 3.	5	15	1. 3.	20
16	1. 2.	32	16	1. 3.	8	16	1. 3.	24	16	2. 0.	0
17	1. 3.	9	17	1. 3.	26	17	2. 0.	3	17	2. 0.	20
18	1. 3.	26	18	2. 0.	4	18	2. 0.	22	18	2. 1.	0
19	2. 0.	3	19	2. 0.	22	19	2. 1.	1	19	2. 1.	20
20	2. 0.	20	20	2. 1.	0	20	2. 1.	20	20	2. 2.	0
21	2. 0.	37	21	2. 1.	18	21	2. 1.	39	21	2. 2.	20
22	2. 1.	14	22	2. 1.	36	22	2. 2.	18	22	2. 3.	0
23	2. 1.	31	23	2. 2.	14	23	2. 2.	37	23	2. 3.	20
24	2. 2.	8	24	2. 2.	32	24	2. 3.	16	24	3. 0.	0
25	2. 2.	25	25	2. 3.	10	25	2. 3.	35	25	3. 0.	20
30	3. 0.	30	30	3. 1.	20	30	3. 2.	10	30	3. 3.	0
40	4. 1.	0	40	4. 2.	0	40	4. 3.	0	40	5. 0.	0
50	5. 1.	10	50	5. 2.	20	50	5. 3.	30	50	6. 1.	0
60	6. 1.	20	60	6. 3.	0	60	7. 0.	20	60	7. 2.	0
70	7. 1.	30	70	7. 3.	20	70	8. 1.	10	70	8. 3.	0
80	8. 2.	0	80	9. 0.	0	80	9. 2.	0	80	10. 0.	0
90	9. 2.	10	90	10. 0.	20	90	10. 2.	30	90	11. 1.	0
100	10. 2.	20	100	11. 1.	0	100	11. 3.	20	100	12. 2.	0
200	21. 1.	0	200	22. 2.	0	200	23. 3	0	200	25. 0.	0

The compt of meafured Land.

	21. Falles				22. Falles.				23. Falles.				24. Falles.	
Fal.		Ro.	Fal.	Fal.		Ro.	Fal.	Fal.		Ro.	Fal.	Fal.		Rood.	Fal.
1			21	1			22	1			23	1			24
2		1.	2	2		1.	4	2		1.	6	2		1.	8
3		1.	23	3		1.	26	3		1.	29	3		1.	32
4		2.	4	4		2.	8	4		2.	12	4		2.	16
5	Acres	2.	25	5	Acres	2.	30	5	Acres	2.	35	5	Acres	3.	0
6		3.	6	6		3.	12	6		3.	18	6		3.	24
7		3.	27	7		3.	34	7	1.	0.	1	1.	0.	8
8	1.	0.	8	8	1.	0.	16	8	1.	0.	24	1.	0.	32
9	1.	0.	29	9	1.	0.	38	9	1.	1.	7	1.	1.	16
10	1.	1.	10	10	1.	1.	20	10	1.	1.	30	1.	2.	0
11	1.	1.	31	11	1.	2.	2	11	1.	2.	13	1.	2.	24
12	1.	2.	12	12	1.	2.	24	12	1.	2.	36	1.	3.	8
13	1.	2.	33	13	1.	3.	6	13	1.	3.	19	1.	3.	32
14	1.	3.	14	14	1.	3.	28	14	2.	0.	2	2.	0.	16
15	1.	3.	35	15	2.	0.	10	15	2.	0.	25	2.	1.	0
16	2.	0.	16	16	2.	0.	32	16	2.	1.	8	2.	1.	24
17	2.	0.	37	17	2.	1.	14	17	2.	1.	31	2.	2.	8
18	2.	1.	18	18	2.	1.	36	18	2.	2.	14	2.	2.	32
19	2.	1.	39	19	2.	2.	18	19	2.	2.	37	2.	3.	16
20	2.	2.	20	20	2.	3.	0	20	2.	3.	20	3.	0.	0
21	2.	3.	1	21	2.	3.	22	21	3.	0.	3	3.	0.	24
22	2.	3.	22	22	3.	0.	4	22	3.	0.	26	3.	1.	8
23	3.	0.	3	23	3.	0.	26	23	3.	1.	9	3.	1.	32
24	3.	0.	24	24	3.	1.	8	24	3.	1.	32	3.	2.	16
25	3.	1.	5	25	3.	1.	30	25	3.	2.	15	3.	3.	0
30	3.	3.	30	30	4.	0.	20	30	4.	1.	10	4.	2.	0
40	5.	1.	0	40	5.	2.	0	40	5.	3.	0	6.	0.	0
50	6.	2.	10	50	6.	3.	20	50	7.	0.	30	7.	2.	0
60	7.	3.	20	60	8.	1.	0	60	8.	2.	20	9.	0.	0
70	9.	0.	30	70	9.	2.	20	70	10.	0.	10	10.	2.	0
80	10.	2.	0	80	11.	0.	0	80	11.	2.	0	12.	0.	0
90	11.	3.	10	90	12.	1.	20	90	12.	3.	30	13.	2.	0
100	13.	0.	20	100	13.	3.	0	100	14.	1.	20	15.	0.	0
200	26.	1.	0	200	27.	2.	0	200	28.	3.	0	30.	0.	0

The compt of measured Land.

Fal.	25. Falles			Fal.	26. Falles			Fal.	27. Falles			Fal.	28. Falles		
	Acres	Rood	Fal.		Acres	Rood	Fa.		Acres	Rood	Fal.		Acres	Rood	Fal.
1.			25	1			26	1			27	1			28
2		1.	10	2		1.	12	2		1.	14	2		1.	16
3		1.	35	3		1.	38	3		2.	1	3		2.	4
4		2.	20	4		2.	24	4		2.	28	4		2.	32
5		3.	5	5		3.	10	5		3.	15	5		3.	20
6		3.	30	6		3.	36	6	1.	0.	2	6	1.	0.	8
7	1.	0.	15	7	1.	0.	22	7	1.	0.	29	7	1.	0.	36
8	1.	1.	0	8	1.	1.	8	8	1.	1.	16	8	1.	1.	24
9	1.	1.	25	9	1.	1.	34	9	1.	2.	3	9	1.	2.	12
10	1.	2.	10	10	1.	2.	20	10	1.	2.	30	10	1.	3.	0
11	1.	2.	35	11	1.	3.	6	11	1.	3.	17	11	1.	3.	28
12	1.	3.	20	12	1.	3.	32	12	2.	0.	4	12	2.	0.	16
13	2.	0.	5	13	2.	0.	18	13	2.	0.	31	13	2.	1.	4
14	2.	0.	30	14	2.	1.	4	14	2.	1.	18	14	2.	1.	32
15	2.	1.	15	15	2.	1.	30	15	2.	2.	5	15	2.	2.	20
16	2.	2.	0	16	2.	2.	16	16	2.	2.	32	16	2.	3.	8
17	2.	2.	25	17	2.	3.	2	17	2.	3.	19	17	2.	3.	36
18	2.	3.	10	18	2.	3.	28	18	3.	0.	6	18	3.	0.	24
19	2.	3.	35	19	3.	0.	14	19	3.	0.	33	19	3.	1.	12
20	3.	0.	20	20	3.	1.	0	20	3.	1.	20	20	3.	2.	0
21	3.	1.	5	21	3.	1.	26	21	3.	2.	7	21	3.	2.	28
22	3.	1.	30	22	3.	2.	12	22	3.	2.	34	22	3.	3.	16
23	3.	2.	15	23	3.	2.	38	23	3.	3.	21	23	4.	0.	4
24	3.	3.	0	24	3.	3.	24	24	4.	0.	8	24	4.	0.	32
25	3.	3.	25	25	4.	0.	10	25	4.	0.	35	25	4.	1.	20
30	4.	2.	30	30	4.	3.	20	30	5.	0.	10	30	5.	1.	0
40	6.	1.	0	40	6.	2.	0	40	6.	3.	0	40	7.	0.	0
50	7.	3.	10	50	8.	0.	20	50	8.	1.	30	50	8.	3.	0
60	9.	1.	20	60	9.	3.	0	60	10.	0.	20	60	10	2.	0
70	10.	3.	30	70	11.	1.	20	70	11.	3.	10	70	12.	1.	0
80	12.	2.	0	80	13.	0.	0	80	13.	2.	0	80	14.	0.	0
90	14.	0.	10	90	14.	2.	20	90	15.	0.	30	90	15.	3.	0
100	15.	2.	20	100	16.	1.	0	100	16.	3.	20	100	17.	2.	0
200	31.	1.	0	200	32.	2.	0	200	33.	3.	0	200	35.	0.	0

G

The compt of meafured Land.

29. Falles.				30. Falles.				40. Falles.		
Fal.		Rood	Fal.	Fal.		Rood	Fal.	Fal.	Rood	Fal.
1			29	1			30	1	1.	0
2		1.	18	2		1.	20	2	2.	0
3		2.	7	3		2.	10	3	3.	0
4	Acres	2.	36	4	Acres	3.	0	4	1. 0.	0
5		3.	25	5		3.	30	5	1. 1.	0
6	1. 0.	14	6	1. 0.	20	6	1. 2.	0
7	1. 1.	3	7	1. 1.	10	7	1. 3.	0
8	1. 1.	32	8	1. 2.	0	8	2. 0.	0
9	1. 2.	21	9	1. 2.	30	9	2. 1.	0
10	1. 3.	10	10	1. 3.	20	10	2. 2.	0
11	1. 3.	39	11	2. 0.	10	11	2. 3.	0
12	2. 0.	28	12	2. 1.	0	12	3. 0.	0
13	2. 1.	17	13	2. 1.	30	13	3. 1.	0
14	2. 2.	6	14	2. 2.	20	14	3. 2.	0
15	2. 2.	35	15	2. 3.	10	15	3. 3.	0
16	2. 3.	24	16	3. 0.	0	16	4. 0.	0
17	3. 0.	13	17	3. 0.	30	17	4. 1.	0
18	3. 1.	2	18	3. 1.	20	18	4. 2.	0
19	3. 1.	31	19	3. 2.	10	19	4. 3.	0
20	3. 2.	20	20	3. 3.	0	20	5. 0.	0
21	2. 3.	9	21	3. 3.	30	21	5. 1.	0
22	3. 3.	38	22	4. 0.	20	22	5. 2.	0
23	4. 0.	27	23	4. 1.	10	23	5. 3.	0
24	4. 1.	16	24	4. 2.	0	24	6. 0.	0
25	4. 2.	5	25	4. 2.	30	25	6. 1.	0
30	5. 1.	30	30	5. 2.	20	30	7. 2.	0
40	7. 1.	0	40	7. 2.	0	40	10. 0	0
50	9. 0.	10	50	9. 1.	20	50	12. 2.	0
60	10. 3.	20	60	11. 1.	0	60	15. 0.	0
70	12. 2.	30	70	13. 0.	20	70	17. 2.	0
80	14. 2.	0	80	15. 0.	0	80	20. 0.	0
90	16. 1.	10	90	16. 3.	20	90	22. 2.	0
100	18. 0.	20	100	18. 3.	0	100	25. 0.	0
200	36. 1.	0	200	37. 2.	0	200	50. 0.	0

The compt of measured Land. 53

	50. Falles.				60. Falles.				70. Falles.				80. Falles		
Fal.	Acres	Ro.	Fal.	Fal.	Acres	Ro.	Fal.	Fal.	Acres	Ro.	Fal.	Fal.	Acres	Ro.	Fal.
1		1	10	1		1.	20	1		1.	30	1		2.	0
2		2.	20	2		3.	0	2		3.	20	2	1.	0	0
3		3.	30	3	1.	0.	20	3	1.	1.	10	3	1.	2	0
4	1.	1.	0	4	1.	2.	0	4	1.	3.	0	4	2.	0.	0
5	1.	2.	10	5	1.	3.	20	5	2.	0.	30	5	2.	2	0
6	1.	3.	20	6	2.	1.	0	6	2.	2.	20	6	3.	0.	0
7	2	0.	30	7	2.	2.	20	7	3.	0.	10	7	3.	2.	0
8	2.	2.	0	8	3.	0.	0	8	3.	2.	0	8	4.	0.	0
9	2.	3.	10	9	3.	1.	20	9	3.	3.	30	9	4.	2.	0
10	3.	0.	20	10	3.	3.	0	10	4.	1.	20	10	5.	0.	0
11	3.	2.	30	11	4.	0.	20	11	4.	3.	10	11	5.	2.	0
12	3.	3.	0	12	4.	2.	0	12	5.	1.	0	12	6.	0.	0
13	4.	0.	20	13	4.	3.	20	13	5.	2.	30	13	6.	2.	0
14	4.	1.	20	14	5.	1.	0	14	6.	0.	20	14	7.	0.	0
15	4.	2.	30	15	5.	2.	20	15	6.	2.	10	15	7.	2.	0
16	5.	0.	0	16	6.	0.	0	16	7.	0.	0	16	8.	0.	0
17	5.	1.	10	17	6.	1.	20	17	7.	1.	30	17	8.	2.	0
18	5.	2.	20	18	6.	3.	0	18	7.	3.	20	18	9.	0.	0
19	5.	3.	30	19	7.	0.	20	19	8.	1.	10	19	9.	2.	0
20	6.	1.	0	20	7.	2.	0	20	8.	3.	0	20	10.	0.	0
21	6.	2.	10	21	7.	3.	20	21	9.	0.	30	21	10.	2.	0
22	6.	3.	20	22	8.	1.	0	22	9.	2.	20	22	11.	0.	0
23	7.	0.	30	23	8.	2.	20	23	10.	0.	10	23	11.	2.	0
24	7.	2.	0	24	9.	0.	0	24	10.	2.	0	24	12.	0.	0
25	7.	3.	10	25	9.	1.	20	25	10.	3.	30	25	12.	2.	0
30	9.	1.	20	30	11.	1.	0	30	13.	0.	20	30	15.	0.	0
40	12.	2.	0	40	15.	0.	0	40	17.	2.	0	40	20.	0.	0
50	15.	2.	20	50	18.	3.	0	50	21.	3.	20	50	25.	0.	0
60	18.	3.	0	60	22.	2.	0	60	26.	1.	0	60	30.	0.	0
70	21.	3.	20	70	26.	1.	0	70	30.	2.	20	70	35.	0.	0
80	25.	0.	0	80	30.	0.	0	80	35.	0.	0	80	40.	0.	0
90	28.	0.	20	90	33.	3.	0	90	39.	1.	20	90	45.	0.	0
100	31.	1.	0	100	37.	2.	0	100	43.	3.	0	100	50.	0.	0
200	62.	2.	0	200	75.	0.	0	200	87.	2.	0	200	100.	0.	0

The compt of measured Land.

	90. Falles.				100. Falles.		
Fal.	Acres	Ro.	Fal.	Fal.	Acres	Ro.	Fal.
1		2.	10	1		2.	20
2	1.	0.	20	2	1.	1.	0
3	1.	2.	30	3	1.	3.	20
4	2.	1.	0	4	2.	2.	0
5	2.	3.	10	5	3.	0.	20
6	3.	1.	20	6	3.	3.	0
7	3.	3.	30	7	4.	1.	20
8	4.	2.	0	8	5.	0.	0
9	5.	0.	10	9	5.	2.	20
10	5.	2.	20	10	6.	1.	0
11	6.	0.	30	11	6.	3.	20
12	6.	3.	0	12	7.	2.	0
13	7.	1.	10	13	8.	0.	20
14	7.	3.	20	14	8.	3	0
15	8.	1.	30	15	9.	1.	20
16	9.	0.	0	16	10.	0.	0
17	9.	2.	10	17	10.	2.	20
18	10.	0.	20	18	11.	1.	0
19	10.	2.	30	19	11.	3.	20
20	11.	1.	0	20	12.	2.	0
21	11.	3.	10	21	13.	0.	20
22	12.	1.	20	22	13.	3.	0
23	12.	3.	30	23	14.	1.	20
24	13.	2.	0	24	15.	0.	0
25	14.	0.	10	25	15.	2.	20
30	16.	3.	20	30	18.	3.	0
40	22.	2.	0	40	25.	0.	0
50	28.	0.	20	50	31.	1.	0
60	33.	3.	0	60	37.	2.	0
70	39.	1.	20	70	43.	3.	0
80	45.	0.	0	80	50.	0.	0
90	50.	2.	20	90	56.	1.	0
100	56.	1.	0	100	62.	2.	0
200	112.	2.	0	200	125.	0.	0

¶ Of building & sclaiting.

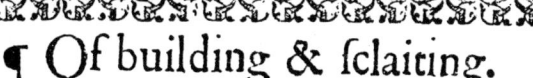

Doe not set downe the manner nor the way, how to measure the Masons nor the Sclaiters workes, because I know not the trew ground and manner thereof, but superceedes that to be done by a common sworne measurer of the best reputation and vnderstanding, who is to measure all workes justly in length and breadth according to some warrand and ground, whereby the owners of the workes knowing the length and breadth of everie House or Wall, Doore or Window, Roofe or Rin-roofe, Storme-window, Ape-house, Easinges, and Windskewes, and all other pieces of worke, they may resort to the Table here following, and finde the just compt what everie particular piece of worke extendeth vnto in Roodes, Elles, and other small quantities, as is set downe in manner following.

A Declaration of the Table

made to finde the quantitie, that ariseth
vpon all worke and labour, that is
measured in length and
breadth.

His Table following is founded vppon the Ell of measure, whereof thirtie six Ells is compted for a Rood of worke, as is said here-tofore, it is set downe in divers Columnes. Each Columne containing three partes: To wit, the breadth of everie piece of worke set downe betweene the two small

black lines that is betweene the end of a Colnmne, and the beginning of another Columne. The length is set downe vppon the left side of everie Columne, and the quantitie that riseth vppon every length and breadth is in the broad roome against the length in the narrow : containing Roodes, Ells, and quarter Ells as they are titled vppon the head of everie number, in the same manner as the former table. Now to finde the compt of everie piece of worke, you shall seeke the compt thereof, as the compt of measured land in the former table.

EXAMPLE OF MASONS WORKE.

A Mason hath builded a wall of 90 Ells of length, and 3 Ells and a halfe Ell of heigth and thicknes : To know what it containes in the whole, you shall cast vp the Columne of 3 Ell, and the Columne of a halfe Ell, which are the Columnes of the breadth, and seeke the length 90 in everie one of them, you will finde in the Columne of 3 Ells against the length 90 standing 7 Roodes 18 Ells, and in the Columne of a halfe Ell against 90 you will finde 1 Rood 9 Ells. These being added together will make 8 Roodes 27 Ells, which is the quantitie of the foresaid wall.

EXAMPLE OF SCLAITING.

A Sclaiter hath theiked a house of length 18 Ells, and of breadth 15 Ells and halfe Ell with 2 Apehouses, everie one of them 3 Ells of length, and of breadth and deepenes compting both the sides 7 Ells: quarter Ell: To know their quantitie, you must seeke everie compt particularly : and first seeke the compt of the house, in casting vp the Columne of 15 Ells of breadth, and the Columne of a halfe Ell, In the Columne of 15 Ells against the length 18, you will finde 7 Roodes 18 Ells, and in the Columne of a halfe Ell against 18 you will finde 9 Ells: These being added will make 7 Roodes 27 Ells. Then to finde the quantitie of the Ape-houses,
seeke

seeke the Columne of 7 Ells, and the Columne of a quarter Ell, which are the breadthes thereof. In the Columne of 7 Ells against the length 3, you will finde 21 Ells. In the Columne of a quarter Ell against 3, you will finde three quarters of a Ell. These will make 21 Ells three quarters of an Ell for each Apehouse. The quantitie of the two Ape-houses, and the quantitie of the house being added together will extend to 8 Roodes, 34 Ells, and a halfe Ell for the quantitie thereof. And such like of all others.

EXAMPLE OF TAPISTRIE.

A Piece of Tapistrie is of length 6 Ells quarter Ell, and 4 Ell halfe Ell of breadth. To knowe the quantitie thereof by this Table, you shall seeke the Columnes of 4 Ells, and the Columne of a halfe Ell, which are the breadth. In the Columne of 4 Ells against the length 6, you will finde 24. Ells, and in the Columne of an halfe Ell, you will finde against 6, standing 3 Ells: Now remember that the small measures of the length must bee reckoned with the breadthe, then seeke the Columne of the small measure of the length, which is of a Quarter Ell, and against 4. you will finde 1. Ell. Now you haue an halfe Ell to bee compted, and multiplyed with the Quarter Ell, which are the small measures to finde their quantitie, resort to this small Table here set downe, and seeke the one of the small measures at the head of the Table, and the other at the side thereof: and where they meet you will finde the quantitie as the Columne of quarter Ell, and the Columne of halfe Ell, you will finde one eight part Ell, or halfe quarter Ell at their meeting. These being added together will extend to 28 Ells and halfe quarter Ell, for the quantitie of the said piece of Tapistrie, and so of all others: as you shall perceiue by this Table in the next page following.

Table for Tapestrie.

	$\frac{1}{4}$	$\frac{1}{2}$	$\frac{3}{4}$
$\frac{1}{4}$	$\frac{1}{16}$	$\frac{1}{8}$	$\frac{3}{16}$
$\frac{1}{2}$	$\frac{1}{8}$	$\frac{1}{4}$	$\frac{3}{8}$
$\frac{3}{4}$	$\frac{3}{16}$	$\frac{3}{8}$	$\frac{9}{16}$

The compt of Building, and Sclating.

El-	Qu. El	Quar		El.	Qu.		El.	Qu.			El.
1		1	5	2.	2	9	6.	3	15		15
2	*Elnes*	2	6	3.	0	10	7.	2	16		16
3		3	7	3.	2	11	8.	1	17		17
4	1.	0	8	4.	0	12	9	0	18	*Roodes*	18
5	1.	1	9	4.	2	13	9.	3	19		19
6	1.	2	10	5.	0	14	10.	2	20		20
7	1.	3	11	5.	2	15.	11.	1	30		30
8	2.	0	12	6.	0	16	12.	0	40	1.	4
9	2.	1	13	6.	2	17	12.	3	50	1.	14
10	2.	2	14	7.	0	18	13.	2	60	1.	24
11	2.	3	15	7.	2	19	14.	1	70	1.	34
12	3.	0	16	8.	0	20	15.	0	80	2.	8
13	3.	1	17	8.	2	30	22.	2	90	2.	18
14	3.	2	18	9.	0	40	30.	0	100	2.	28
15	3.	3	19	9.	2	50	1. 1. 2			2. *Elnes.*	
16	4.	0	20	10.	0	60	1. 9. 0		1		2
17	4.	1	30	15.	0	70	1. 16. 2		2		4
18	4.	2	40	*Roodes* 20.	0	80	1. 24. 0		3		6
19	4.	3	50	25.	0	90	1. 31. 2		4		8
20	5.	0	60	30.	0	100	2. 3. 0		5		10
30	7.	2	70	0. 35.	0	*Elne*	*An Elne.*		6		12
40	10	0	80	1. 4.	0	1			1 7		14
50	12.	2	90	1. 9.	0	2			2 8.		16
60	15.	0	100	1. 14.	0	3			3 9		18
70	17.	2		3 of *Elne.*		4			4 10		20
80	20.	0		4		5			5 11		22
90	22.	2	*Elne*			6			6 12	*Rood*	24
100	25.	0	1	*Elnes*	3	7			7 13		26
			2	1.	2	8			8 14		28
	Halfe Elne.		3	2.	1	9			9 15		30
			4.	3.	0	10			10 16		32
1			2 5.	3.	3	11			11 17		34
2	1.	0	6	4.	2	12			12 18	1.	0
3	1.	2	7	5.	1	13			13 19	1.	2
4	2.	0	8	6.	0	14			14 20	1.	4

The compt of Building, and Sclaiting.

El	Rood	El.		El.			El.		Rood	El.		Rood	E l.
30	1.	24	90	7.	18	4		20	14	2.	12		32
40	2.	8	100	8.	12	5		25	15	2.	18	5	4
50	2.	28		4. Elnes.	6			30	16	2.	24	5 1.	12
60	3.	12			4	7		35	17	2.	30	7 1.	20
70	3.	32	1 2		8	8	1.	4	18	3.	0	8 1.	28
80	4.	16			12	9	1.	9	19	3.	6	9 2.	0
90	5.	0	3		16	10	1.	14	20	3.	12	10 2.	8
100	5.	20	4		20	11	1.	19	21	3.	18	11 2.	16
	3. Elnes.		5		24	12	1.	24	22	3.	24	12 2.	24
			6		28	13	1.	29	23	3.	30	13 2.	32
1.		3	7		32	14	1.	34	24	4.	0	14 3.	4
2.		6	8									15 3.	12
3		9	9	1.	0	15	2.	3		7. Elnes.		16 3.	20
4		12	10	1.	4	16	2.	8	1.		7	17 3.	28
5		15	11	1.	8	17	2.	13	2.		14	18 4.	0
6		18	12	1.	12	18	2.	18	3.		21	19 4.	8
7		21	13	1.	16	19	2.	23	4.	Roodes	28	20 4.	16
8		24	14	1.	20	20	2.	28	5.		35	21 4.	24
9		27	15	1.	24	21	2.	33	6.		6	22 4.	32
10	Roodes	30	16	1.	28	22	3.	2	1. 7.		13	23 5.	4
11		33	17	1.	32	23	3.	7	1.		20	24 5.	12
12	1.	0	18	2.	0	24	3.	12	1.		27		
13	1.	3	19	2.	4		6. Elnes.		1.		34	9. Elnes	
14	1.	6	20	2.	8	1		6	2.		5		
15	1.	9	30	3.	12	2		12	2.		12	1	9
16	1.	12	40	4.	16	3		18	2.		19	2 Roodes	18
17	1.	15	50	5.	20	4	Roodes	24	2.		26	3	27
18	1.	18	60	6.	24	5		30	2.		33	4 1.	0
19	1.	21	70	7.	28	6	1.	0	3.		4	5 1.	9
20	1.	24	80	8.	32	7	1.	6	3.		11	6 1.	18
30	2.	18	90	10.	0	8	1.	12	3.		18	7 1.	27
40	3.	12	100	11.	4	9	1.	18	3.		25	8 2.	0
50	4.	6		5. Elnes.		10	1.	24	3.		32	9 2.	9
60	5.	0	1		5	11	1.	30	4.		3	10 2.	18
70	5.	30	2		10	12	2.	0	4.		10	11 2.	27
80	6.	24	3		15	13	2.	6	4.		17	12 3.	0
									4.		24	13 3.	9
										8. Elnes.		14 3	18
											8	15 3.	27
									1.		16	16 4.	0
									2.		24	17 4.	9
									3.			18 4.	18

Of Building and Sclaiting.   61

Eln.	Rood	El	Eln.	Rood	El	Eln.	Rood	El	Eln.	Rood	El	Eln.	Rood	El
19	4.	27			3			33	13	4.	12	23	8.	1
20	5.	0	1.		8			14	4.	24	24	8.	24	
21	5.	9	1.		19			15	5.	0		14. Elne.		
22	5.	18	1.		30			16	5.	12			14	
23	5.	27	2.		5			17	5.	24	1		28	
24	6.	0	2.		16			18	6.	0	2	1.	6	
		9	2.		27			19	6.	12	3	1.	20	
	10. Elne		3.		2			20	6.	24	4	1.	34	
1		10	3.		13			21	7.	0	5	1.	34	
2		20	3.		24			22	7.	12	6	2.	12	
3		30	3.		35			23	7.	24	7	2.	26	
4	1.	4	4.		10			24	8.	0	8	3.	4	
5	1.	14	4.		21					9			18	
6	1.	24	4.		32			13. Elne				3.	18	
7	1.	34	5.		7				13	1		3.	32	
8	2.	8	5.		18				26	2		4.	10	
9	2.	18	5.		29	3	1.	3	13	5.	2			
10	2.	28	6.		4	4	1.	16	14	5.	16			
11	3.	2	6.		15	5	1.	29	15	5.	30			
12	3.	12	6.		26	6	2.	6	16	6.	8			
13	3.	22	7.		1	7	2.	19	17	6.	22	1		16
14	3.	32	7.		12	8	2.	32	18	7.	0	2		32
15	4.	6		12. Elne		9	3.	9	19	7.	14	3	1.	12
16	4.	16			10	10	3.	22	20	7.	28	4	1.	28
17	4.	26			11	11	3.	35	21	8.	6	5	2.	8
18	5.	0			12	12	4.	12	22	8.	20	6	2.	24
19	5.	10	1.		13	13	4.	25	23	8.	34	7	3.	4
20	5.	20	1.		14	14	5.	2	24	9.	12	8	3.	20
21	5.	30	1.		15	15	5.	15				9	4.	0
22	6.	4	2.		16	16	5.	28		15. Elne		10	4.	16
23	6.	14	2.		17	17	6.	5	1		15	11	4.	32
24	6.	24	2.		18	18	6.	18	2		30	12	5.	12
		9	3.		19	19	6.	31	3	1.	9	13	5.	28
	11. Elne		3.		20	20	7.	8	4	1.	24	14	6.	8
1		11	3.		21	21	7.	21	5	2.	3	15	6.	24
2		22	4.		22	22	7.	34	6	2.	18	16	7.	4

Of Building and Sclaiting.

Eln.	Rood.	E	Elne	Rood.	El	Elne	Rood.	El	Elne	Rood.	El	Elne	Rood El.
17	7.	20	2	1.	0	13	6.	31	24	13	12	10	6. 4
18	8.	0	3	1.	18	4	7.	14		21. Elne		11	6. 26
19	8.	16	4	2.	0	15	7.	33	1		21	12	7. 12
20	8.	32	5	2.	18	16	8.	16	2	1.	6	13	7. 34
21	9.	12	5	3.	0	17	8.	35	3	1.	27	14	8. 20
22	9.	28	7	3.	18	18	9.	18	4	2.	12	15	9. 6
23	10.	8	8	4.	0	19	10.	1	5	2.	33	16	9. 28
24	10	24	9	4.	18	20	10.	20	6	3.	18	17	10. 14
	17 Elne	0	5.	0	21	11.	3	7	4.	3	18	11. 0
			11	5.	18	22	11.	22	8	4.	24	19	11. 22
1		17	12	6.	0	23	12.	5	9	5.	9	20	12. 8
2		34	13	6.	18	24	12.	24	10	5.	30	21	12. 30
3	1.	15	14	7.	0		20. Elne	11	6.	15	22	13. 16
4	1.	32	15	7.	18	1		20	12	7.	0	23	14. 2
5	2.	13	16	8.	0	1	1.	4	13	7.	21	24	14. 24
6	2.	30	17	8.	18	2	1.	24	14	8.	6		23. Elne
7	3.	11	18	9.	0	3	2.	8	15	8.	27		23
8	3.	28	19	9.	18	4	2.	28	16	9.	12	1	1. 10
9	4.	9	20	10.	0	5	3.	12	17	9	33	2	1. 33
10	4.	26	21	10.	18	6	3.	32	18	10.	18	4	2. 20
11	5.	7	22	11.	0	7	4.	16	19	11.	3	5	3. 7
12	5.	24	23	11.	18	8	5.	0	20	11.	24	6	3. 30
13	6.	5	24	12.	0	10	5.	20	21	12.	9	7	4. 17
14	6.	22		19. Elne	11	6.	4	22	12.	30	8	5. 4
15	7.	3	1		19	12	6.	24	23	13	15	10	5. 27
16	7.	20	1	1.	2	13	7.	8	24	14	0	11	6. 14
17	8.	1	2	1.	21	14	7.	28		22 Elne	12	7. 1
18	8.	18	3	2.	4	15	8.	12	1		22	13	7. 24
19	8.	35	4	2.	23	16	8.	32	2	1.	8	14	8. 11
20	9.	16	5	3.	6	17	9.	16	3	1.	30	15	8. 34
1	9.	33	6	3.	25	18	10.	0	4	2.	16	16	9. 21
22	10.	14	7	4.	8	19	10.	20	5	3.	2	17	10. 8
23	10.	31	8	4.	27	20	11.	4	6	3.	24	18	10. 31
24	11.	12	9	5.	10	21	11.	24	7	4.	10	19	11. 18
	18. Elne	10	5.	29	22	12.	8	8	4.	32	20	12. 5
			11	6.	12	23	12.	28	9	5.	18	21	12. 28
												22	13. 15
												23	14. 2
												24	14. 25
													15. 12

THE DESCRIPTION OF THE
TABLE SET DOWNE TO KNOW
THE WEIGHT OF EVERIE
loafe of wheat bread at all
prices of wheate.

HE particular weight of all quantities of wheate bread is most needefull to bee knowne of all in-dwellers within Brughes. And therefore I haue set downe this Table here following, which I made in Anno 1597 at the desire of the Bailies of Edinburgh, to resolue them what euerie loafe of wheate bread should weigh, at all prices of wheate. It is founded vppon a triall made by the counsell of the said Burgh in Anno 1555: who (after good consideration of the labour and all charges needefull to bee allowed and deduced to the Bakers:) concluded that there should bee made 140 poundes weight of very fine wheate bread out of euerie Bow of wheate. The which Table I amended and omitted forth the price of the wheate and bread not needfull, and haue augmented the prices of wheate betweene 16. lib. and 20 lib. the Bow, with the weight of the 2 shilling loafe, which was not before. This Table, is deuided in these 2 pages following, containing 4 Columnes in euerie page. The first page hath the Columne of the prices of the wheate, beginning at 4 lib. descending downe to 4 lib. 10 shillings. Next to 5 lib. and so foorth to 20 lib. The other 3 Columnes are the weight of the 2 shilling loafe. The Columne of the 18 pennie loafe, and of the 16 pennie loafe, and euerie one of them containes 4 numbers.

K

The firſt are pound weights, the ſecond are ounces, the third are drop weigths, and the 4 number are graine weights, as they are titled and marked vpon the head of everie number, as for partes of graines they are not needfull to bee ſet downe. The ſecond page, hath in like manner the prices of wheate in the firſt Columne, in the ſecond, the weight of the 12 pennie loafe, next of the 8 pennie loafe, and of the 6 pennie loafe with their ſeverall numbers of weight. If any Arithmetician bee curious, to know the partes of graines not ſet downe, let them reſort to me, and I ſhall giue them contentment.

TO FINDE THE WEIGHT OF BREAD by ſome oxamples.

IN caſe the Provoſt, Bailies and Counſell of Edinburgh, after tryall of the markets of Edinburgh, Hadingtoun, and Dalkeith, haue ordained that the Bakers ſhall baike 12 pennie loafes, and to keepe the poiſe or weight according to 13 lib. the Bow of wheate: to know by the Table what weight the ſaid loafe ſhould weigh, you ſhall ſeeke the price 13 lib. in the firſt Columne of the ſecond page, and there againſt it, you will finde 8 ounces 9 drop weight, and 30 graines for the weight thereof. Another example. The Bakers are ordained to baike 16 pennie loafes, according to 12 lib. 10 ſhillings the Bow of wheate: To finde the weight thereof, you will finde the Columne of the ſaid loafe in the firſt page, and ſeeke the price of 12 lib. 10 ſhillings in the firſt Columne, and goe forth in one line towards the right hand, and you will finde in that Columne, againſt the ſaid price 11 ounces 15 drop weight and 5 graines: The thrid example. The Bakers are ordained to baike 18 pennie loafes, according to 10 lib. 10 ſhillinges the Bow of wheate, you will finde in the Columne of 18 pennie bread, againſt 10 lib. 10 ſhillings, 16 ounces for the weight thereof, and ſo forth of all other bread.

A TABLE to finde out

PRYCES of Wheat.	The 2 shilling Bread				The 18 penny loafe				The 16 penny loafe.			
	pund	vnce	drops	grains	punds	vnces	drops	grains	punds	vnces	drops	graines
4. lib.	3	8			2	10			2	5	5	12
4. lib. 10. shil.	3	1	12	16	2	5	5	12	2	1	2	34
5. lib.	2	12	12	28	2	1	9	21	1	13	13	31
5. lib. 10 shil.	2	8	11	22	1	14	8	26	1	11	2	15
6. lib.	2	5	5	12	1	12			1	8	14	8
6. lib. 10. shil.	2	2	7	13	1	9	13	19	1	6	15	21
7. lib.	2				1	8			1	5	5	12
7. lib. 10. shil.	1	13	13	31	1	6	6	14	1	3	14	20
8. lib.	1	12			1	5			1	2	10	24
8. lib. 10. shil.	1	10	5	23	1	3	12	8	1	1	9	3
9. lib.	1	8	14	8	1	2	10	24		16	9	17
9. lib 10. shil.	1	7	9	9	1	1	10	34		15	11	18
10. lib.	1	6	6	14		16	12	28		14	14	33
10. lib. 10. shil.	1	5	5	12		16				14	3	20
11. lib.	1	4	5	29		15	4	13		13	9	7
11. lib. 10. shil.	1	3	7	23		14	9	26		12	15	27
12. lib.	1	2	10	24		14				12	7	4
12. lib. 10. shil.	1	1	14	25		13	7	1		11	15	5
13. lib.	1	1	3	24		12	14	27		11	7	28
13. lib. 10. shil.		16	9	17		12	7	4		11	0	35
14. lib.		16				12				10	10	24
14. lib. 10. shil.		15	7	6		11	9	13		10	4	28
15. lib.		14	14	33		11	3	7		9	15	10
15. lib. 10 shil.		14	7	8		10	13	15		9	10	5
16. lib.		14				10	8			9	5	12
16. lib. 10. shil.		13	9	7		10	2	32		9	0	29
17. lib.		13	2	29		9	14	4		8	12	18
17. lib. 10. shil.		12	12	18		9	9	21		8	8	19
18. lib.		12	7	4		9	5	12		8	4	26
18. lib. 10 shil.		12	1	26		9	1	10		8	1	5
19. lib.		11	12	22		8	13	17		7	3	27
19. lib. 10. shil.		11	7	28		8	9	30		7	10	19
20. lib.		11	3	7		8	6	14		7	7	16

the weight of Wheat Bread.

PRYCES of Wheat.	The 12. penny loafe				The 8. penny loafe				The 6. penny loafe			
	punds	vnces	drops	grains	punds	vnces	drops	grains	punds	vnces	drops	grains
4. lib.	1	12			1	2	10	24		14		
4.lib.10.shil.	1	8	14	8		16	9	17		12	7	4
5. lib.	1	6	6	14		14	14	33		11	3	7
5.lib.10.shil	1	4	5	29		13	9	7		10	2	7
6. lib.	1	2	10	24		12	7	4		9	5	32
6.lib.10.shil.	1	1	3	24		11	7	28		8	9	12
7. lib.		16				10	10	24		8		30
7.lib.10.shil.		14	14	33		9	15	10		7	7	16
8. lib.		14				9	5	12		7		
8.lib.10.shil.		13	2	29		8	12	19		6	6	14
9. lib.		12	7	4		8	4	26		6	3	20
9.lib.10 shil		11	12	22		7	13	27		5	14	11
10. lib.		11	3	7		7	7	16		5	9	21
10.lib.10.shil		10	10	24		7	1	28		5	5	12
11. lib.		10	2	32		6	12	21		5	1	16
11.lib.10.shil		9	11	29		6	7	31		4	13	32
12. lib		9	5	12		6	3	20		4	10	24
12.lib.10.shil		8	15	12		5	15	20		4	7	24
13. lib.		8	9	30		5	11	32		4	4	33
13.lib.10 shi		8	4	26		5	8	17		4	2	13
14. lib.		8				5	5	12		4		
14.lib.10 shil		7	11	21		5	2	14		3	13	23
15. lb.		7	7	16		4	15	23		3	11	26
15.lib.10 shil		7	3	22		4	13	2		3	9	29
16. lib.		7				4	10	24		3	8	0
16.lib.10 shi		6	12	21		4	8	14		3	6	10
17. lib.		6	9	14		4	6	9		3	4	25
17.lib.10 shil		6	5	14		4	4	9		3	3	7
18. lib.		6	3	20		4	2	13		3	1	28
18.lib.10 shil		6	0	31		4	0	20		3	0	15
19. lib.		5	14	11		3	14	31		2	15	5
19.lib.10 shil		5	11	32		3	13	9		2	13	34
20. lib.		5	9	21		3	11	26		2	12	28

THE preceeding Table is founded but vppon 140 pound weight of fine wheate bread: to bee made of everie Bow of wheate, conforme to the tryall made by the Counsell of Edinburgh, and ordinance set downe there-vpon in Anno 1555 as said is. But now the said Counsell finding that albeit some of the Bakers makes better bread then the rest: yet the best bread is not of that finenesse, that was ordained by that ordinance: and therefore are of intention to make new trialls: like as the Burrowes at their meeting in Aberdene, appointed the same to bee done at sundrie Burghes, for trying of all sort of wheate: the which trialls being made and reported, I thinke that they will finde that the Bowe of wheate, may render a greater quantitie of bread, then is set downe in the said ordinance. And because all Lieges may not eate of one kinde of bread, nor yet should drinke of a like sort of drinke, they will not onely make triall vppon the wheat, which may render two sortes of bread, but also of the Rye, Oates, Beanes, and Pease, for course bread to the meanest sort. And then the prices of victuall being modefied after the rate of the fore-said markets, by the said Counsell, and set in write vppon the crosse monethlie, conforme to their ancient forme: to informe the Lieges of the prices of victuall monethlie, the Tables to be made conforme to the new trialls, will shew them what weight of bread they should haue for their money, conforme to the modefied price of victuall: and so all persons will bee controllers of the poise and weight of bread, to ease the Magistrates, and make the Lieges to bee more dewlie vsed. I doe thinke they are also of intention to make triall vppon the Beere and Malt, to trie what number of gallons of double and single Ale and Beere the Bow of Malt may render: and thereby to finde out the price of the pinte, both of Ale and Beere.

    I was of intention to haue set downe the Weightes, Metts, Measures, and coynes of all our neighbour countries, with the difference betweene them and this Nation in everie thing: but I will omit that and other thinges, vntill I heare how this will bee accepted, hoping that the best sort will take in good part my honest meaning.

All praise to God.

QC
89
G82
S354
1974

SEP 16 1975